giving
nature
a home

Otters

Nicola Chester

giving
nature
a home

The RSPB is the country's largest nature conservation charity,
inspiring everyone to give nature a home so that birds and wildlife can thrive again.

By buying this book you are helping to fund The RSPB's conservation work.

If you would like to know more about The RSPB, visit the website at www.rspb.org.uk
or write to: The RSPB, The Lodge, Sandy, Bedfordshire, SG19 2DL; 01767 680551.

First published in 2014

Bloomsbury Publishing Plc, 50 Bedford Square, London WC1B 3DP
Bloomsbury USA, 175 Fifth Avenue, New York, NY 10010

www.bloomsbury.com
www.bloomsburyusa.com

Bloomsbury Publishing, London, New Delhi, New York and Sydney

A CIP catalogue record for this book is available from the British Library
Library of Congress Cataloging-in-Publication Data has been applied for

Commissioning editor: Julie Bailey
Project editor: Alice Ward
Series design and layout: Rod Teasdale

ISBN (print) 978-1-4729-0386-0
ISBN (Ebook) 978-1-4729-0387-7

Printed in China by C&C Offset Printing Co Ltd.

10 9 8 7 6 5 4 3 2 1

MIX
Paper from
responsible sources
FSC® C008047

For my own cubs, Billy, Evie and Rosie, and for Martin.
Remember, frogs are not ice cream.

Contents

Meet the Otters   4

Otters Around the World   18

Food and Home: a Riverine Life   30

Agility and Grace: a Life in the Water   40

Family and Play   50

Almost Extinct: the Otter's Past   62

Threats and Recovery   72

Otter Spotting   86

Salty Dogs: an Adaptation   104

The Otter in Writing   112

Glossary   124

Further Reading   124

Resources   125

Tailpiece   127

Acknowledgements   127

Index   128

Meet the Otters

Otters are among the most exciting – and elusive – mammals living alongside us today. They are agile, intelligent and playful, and many of us dream of seeing one in the wild. Otters are so well adapted to their environment that when they enter the water they simply pour themselves in, soundlessly melting away and becoming part of the river or sea itself – another ripple in the current or a chain of bubbles bursting on the surface.

Our name for these elemental creatures has evolved from our old word *wodr*, from which we also get the word water, hinting at a time when Otters were more prevalent (though no more conspicuous), and when their sinuous slip of a spirit seemed indistinguishable from the element of water itself. They are almost mythical, and made more exciting because of course they are not. Yet often all you see for proof that they exist are hints, signs that they have been there: a half-eaten fish on the riverbank, a sweet-smelling, fishy poo or pawprints like little sunrays in the silty beach at a bend of a river.

   As a predator at the very top of a food chain, the Otter was never numerous, but its presence – or the hint of its presence (the merest sign one has passed through) – is the Holy Grail of river conservation, the gold stamp indicating the health of a river and all that lives in and around it. An Otter in the water produces a skip-a-heartbeat moment of joy for any conservationist, and a hint that we may be getting things right in living with nature.

Above: Elusive, lithe, playful, and able to shift so effortlessly from land to water, Otters cast a spell of an elemental kind.

Opposite: Seeing a wild otter is thrilling, but their presence also indicates an environment healthy enough to give life support to a whole ecosystem.

Water dogs

Our Otter is the Eurasian Otter (*Lutra lutra*) and the only otter species found in Britain and continental Europe. It is also known as the European River Otter and the Old World Otter. Otters are secretive, mostly nocturnal and semi-aquatic – they are never far from a source of water. They are sleek, lithe, streamlined animals with an exuberance about them that suggests boundless energy. All members of the otter family are dark chocolate-brown, often with a pale creamy chin, throat and belly. They have bright, intelligent eyes, small ears and lots of marvellous, highly sensitive whiskers.

Below: An exuberant leap into the Little Ouse, Norfolk: perfectly streamlined and adapted for life in water, Otters are never far from it and seem to revel in their swimming and diving ability.

Otters belong to the Mustelidae or weasel family of carnivorous (meat-eating) mammals. Like all mustelids, they are long-bodied with short, powerful legs, and feet with five clawed toes. However, unlike other mustelids, Otters have webbed feet and long, thick, muscular tails – both of these attributes help them move efficiently through the water.

Otters are surprisingly large, at over a metre (3¼ft) from the nose to the tip of the Labrador-like tail, and they are often described as dog-like. Male Otters are in fact known as 'dogs' and females as 'bitches'. Although their young are known as cubs, they often get called pups or even kits, but never puppies. In Welsh they are called *dŵr-gi*, which means water dog, and in Scottish Gaelic,

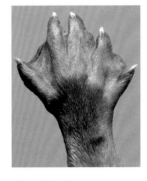

Above: The only serious swimmer in the family of British mustelids; webs of skin stretch between the Otter's five claws, helping to propel it through the water.

MEET THE OTTERS

Above: With fluid lines and such a long, low profile, otters can melt from land into water with barely a sound or a ripple.

Right: An otter takes a gulp of air before diving. At a quick glance, a swimming otter could be mistaken for a doggy-paddling Border Terrier. Most of their surprisingly large body is submerged.

matadh, meaning hound. In the Irish language they are called *dobhar-chù*, which again means water dog.

An Otter swims low in the water, with just its eyes, nostrils and the top of its broad, flat head and part of its long back visible. Its tapering, muscular tail acts like a rudder and can be half the length of its body. When an Otter is cruising along, its partly submerged head can make it look a little like a swimming Border Terrier, but Otters also dive and break the surface of the water in the manner of dolphins, arching their backs and lifting their tails – yet slipping beneath the water with barely a splash. Sometimes their presence underwater is only revealed by a glimpse of a tail or a line of bubbles rising to the surface of the water. Swimming Otters can be confused with Mink, animals you are probably more likely to see. But at about half the size of Otters, mink are much smaller, and they are fluffier, more buoyant in the water and have very pointy, ferret-like faces.

On land Otters really can look like dogs and will bound along the water's edge or across roads between

habitats with a hump-backed gambol. When they are wet they look black and their fur is spiky. Like other mustelids, Otters can stand up straight on their hind legs, balancing on their long back feet and tail, forepaws hanging down like (unrelated) Meerkats, to look around.

Otters are mostly silent, but do communicate with whistles, chirrups and chattering, twittery noises.

Living alongside a body of water, from rivers, lakes and canals to coastal waters, Otters eat mostly fish. However, they are also opportunists and will readily turn to frogs, waterbirds, crabs and even rabbits. Their innate inquisitiveness helps them search out, explore and exploit new food sources, which can bring them into contact – and conflict – with man.

Male Otters are larger and heavier than females. An average adult male's head and body measures approximately 80cm (31in) with a 40cm (16in) tail, while an average female's head and body measures around 70cm (28in) with a 35cm (14in) tail. Male Otters weigh approximately 10kg (22lb), and females average about 7kg (15lb).

Otters can live for up to 10 years, although their lives are tough, with many hazards, and few survive for more than five years.

Left: Being low to the ground could be a disadvantage, but otters frequently stand up, Meerkat-style, balancing on the 'flippers' of their big hindfeet and supported by the base of a thick tail.

Below: Although Otters are solitary animals, they form bonds with family members that are sometimes renewed on meeting with affectionate play.

A scientific family tree

There are 13 species of otter worldwide. Otters are part of the Carnivora, a large order of carnivorous mammals, and the Mustelidae are the largest family in that order. Mustelids are descended from prehistoric wolf-like creatures called vulperines. They have been around for a long time, along with the rodent family (mice, voles, rats and so on). They are thought to be one of the oldest families of carnivore, first appearing about 40 million years ago. Otters are one of the earliest carnivores and are descended from animals called *Mionictis* that lived 30 million years ago. The mustelids we are familiar with today have directly evolved from animals that appeared about 15 million years ago. By contrast, we modern humans began to split away and evolve from our common ape-like ancestor between five and seven million years ago. We are relative newcomers; Otters have been around a lot longer than us.

Below: One of the earliest carnivores, Otters are part of the mustelid (or weasel) family and have evolved over 30million years or more to exploit the niche of their entirely watery habitat. Humans are comparative newcomers.

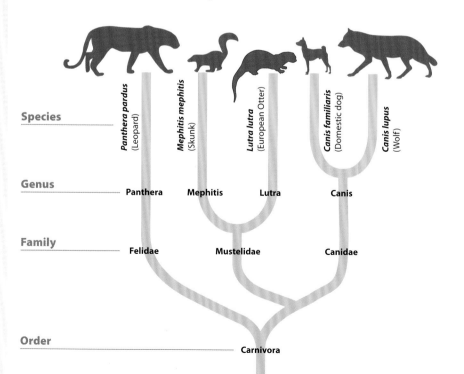

Species

Panthera pardus (Leopard)

Mephitis mephitis (Skunk)

Lutra lutra (European Otter)

Canis familiaris (Domestic dog)

Canis lupus (Wolf)

Genus   Panthera   Mephitis   Lutra   Canis

Family   Felidae   Mustelidae   Canidae

Order   Carnivora

Long body, short legs, and a scent to remember

From the smallest (our native British Weasels are just 20cm/7¾in long and weigh approximately 50g/1¾oz) to the largest (South American Giant Otters are on average 2m/6½ft long and weigh 30kg/66lb), mustelids all have certain things in common. They have a long body shape, short legs and thick fur, and feet with five toes and dog-like, non-retractable claws (unlike those of cats). They tend to have small but broad, flattened skulls and short, rounded ears. Mustelids have well-developed anal scent glands that they use for marking territory and messaging. Indeed, when tracking or recording the presence of Otters, this trait comes in very handy for us humans, too!

Above: The undulating, hump-backed profile of an otter carrying off its catch. Short, powerful legs and a low slung, elongated back are the hallmarks of a mustelid.

Many of the old names for mustelids come from this pungent trait. Foulmart is the old name for the Polecat, but sweetmart is the name for the nicer smelling Pine Marten; its poo does not quite smell of roses, but many people liken it to Parma violets! The domesticated ferret's scientific name is *Mustela Putorius furo*, which translates as putrid thief.

Most mustelids are solitary, nocturnal and do not hibernate, staying active year-round. They all have a high metabolism (the speed at which they digest and convert food to energy) and therefore must spend a lot of time foraging, hunting and chasing down prey, then recovering.

Otters worldwide

Mustelids (and indeed otters) can be found all over the world, with the exception of Australasia and Antarctica. Perhaps because of their need to constantly seek out food, they have evolved into curious, intelligent animals that can climb, dig, stand on their hind legs, swim and run fast. Some mustelids, like the Pine Marten, are arboreal (living in trees), whereas others, like the Badger, spend much of their lives underground where they dig a complex network of tunnels and chambers, known as setts. As a measure of their intelligence, otters have learned to exploit opportunities created by man and will explore fishing nets and fisheries. Sea Otters have even evolved to use rocks as tools to get at their food, one of the few mammal species to do so. Ferrets, the domesticated mustelids closely related to the Polecat, are kept as pets or taught to hunt rats and Rabbits, sometimes alongside a trained dog, hawk or falcon.

Many of the mustelids are much stronger than their size might indicate and will kill and eat prey much larger than themselves. A Stoat can kill a Rabbit more than twice its size, and Wolverines are aggressive Arctic and tundra scavengers that can tear frozen meat from a carcass and crush through the thigh bones of a moose, an animal five times bigger than themselves, to get at the marrow.

Below: A Sea Otter floats off Monterey Bay, California. Sea Otters have evolved to live almost entirely at sea, and are the only species of otter not to have the scent glands all other mustelids have.

The British mustelids

Most of Britain's wild carnivores are mustelids, and although some are fairly common they can be difficult to see in the wild.

- **Otter** The only truly amphibious mustelid, living an almost exclusively watery life. The Otter has lived in the British Isles for at least the last 10,000 years, since the last ice age. It is an excellent, graceful and energetic swimmer, chasing and catching much of its fishy prey in rivers, streams, lakes and some areas by the sea.

- **Badger** The largest mustelid in Britain. A widespread, familiar black-and-white nocturnal animal that lives in family groups in setts below the ground. Its staple diet of earthworms is supplemented by whatever else comes its way.

- **Weasel** The smallest mustelid in the world. Looking like a stretched mouse, it is a fierce and energetic predator capable of pursuing voles, mice and rats down small tunnels. It is chestnut coloured with a short tail and a creamy throat and belly. The Weasel is widespread, but absent from Ireland.

- **Stoat** Similar to the Weasel in appearance and colouration, yet larger and always with a sooty black tip to its tail, even in some animals that turn white in winter. It hunts larger prey such as Rabbits and birds. Stoats occur throughout Britain and Ireland.

- **Polecat** Larger than a stoat, with very dark brown fur that appears black, and yellowish-brown underfur. It has a distinctive 'bandit' face-mask and white ear-tips. The Polecat has a diet similar to that of the Stoat. It is making a comeback from near extinction.

- **Pine Marten** Tree-climbing mustelid with thick chocolate-brown fur, cream or orangey throat-patches, and a long, fluffy tail. Its diet includes mice, birds and their eggs, beetles, berries and fungi. Pine Martens are found in Scotland and Ireland, but are now rare in England and much of Wales.

- **Mink** The North American Mink was farmed for its fur, and after escaping or being released is now widespread. However, it is highly invasive and threatens our native wildlife – particularly the Water Vole – and attempts are ongoing to eradicate it. The mink is dark chocolate-brown with a white chin, chest and belly. It is a good swimmer and although much smaller than the Otter, with a sharp, pointy face and fluffy tail, it is the most likely of animals to be confused with one, particularly when swimming.

- **Polecat-ferret** The offspring of an escaped feral (gone wild) ferret that has cross-bred with a Polecat. It has lighter coloured fur than a Polecat and a paler face. Like the Polecat, its diet is also very similar to the stoat's.

An encounter with Otters

We wade through waist-high grasses and nettles up to our ears. A heron launches from the sallow trees on great, creaky wings like Old Nog in the opening lines of *Tarka the Otter*. I'd have taken it as a good sign, if I thought I'd be that lucky.

Frothy Meadowsweet and Brooklime grow from the river margins of this private chalk stream we have been given permission to walk along on Thursday evenings, when no one fishes. Mid-river, thick blue-green reeds bow and quiver as if something other than the current is moving them from below.

Twice, we hear the distinct 'plop' of Water Voles entering the water. Sedge Warblers bustle in the tall ranks of reeds, climbing the stems like little mice.

We find Otter-sized tunnels through the grass and a spraint placed prominently on top of a tump of great Tussock Sedge, bleached by marking. Above the Comfrey dozens of bright Scarlet Tiger Moths flutter among drabber Meadow Brown butterflies.

The sun dips and the Swallows descend, twittering and swooping on the fly life that the fish are rising to. Everything changes in this hour, and soon the birds, Scarlet Tigers and butterflies are all gone.

In the time it took us to walk upriver into the sunset and back, the Otters have been out. The grass is still warm and flattened in one place and wet in another, still pricking back up from the press of a large animal.

My heart races. Maybe this time, maybe tonight. I long for whatever they will give away. I have seen a disappearing tail here and in the nearby town, the low, bumpy silhouette of an Otter passing through the sodium-orange pool of a reflected street lamp at dusk.

Then suddenly, out on the bank ahead, its back end concealed by rushes and reeds, the head, neck and front paws of an Otter. We freeze. It moves out and begins to roll on the bank, morphing into two Otters. They roll and tangle together for a moment, one arching its two-tone chocolate-and-cream neck to bite the face of the

Above: Watching wild Otters unawares is an unforgettable experience, whether a chance encounter or the hard-won reward for many patient, uncomfortable hours spent tracking and waiting.

other before both disappear back into the water. And then the soundless sense of something beside me turns my head and there, right there in the river, is the broad, flat head of a swimming Otter. Its nose, eyes and small ears just breach the steady line of the flow, and the wet fur on the nape of its neck is spiky and dark. I turn to silently, frantically, alert my companion, but when I look back it is gone.

When we reach the place where the near-grown cubs came out, I kneel to lay my hand in the place where they rolled in the grass and leave it there for a moment. It is still wet with the water wrung from their pelts. When I look up, their presence has ignited this magical, riverine landscape with an intensity I'd not thought possible.

River Kennet, West Berkshire, Summer 2013

Otters Around the World

Nearly all of the world's 13 otter species are in decline or even threatened with extinction. Their decline is directly linked with our need for water – with how we use it and the land around it. Otters are in decline mostly because of habitat loss and water pollution from industry or farming. They are also hunted (often for their fur or for sport) or killed when coming into conflict with fishermen. Many are killed on roads.

Otters' ambassador

The Asian Small-clawed Otter (*Aonyx cinereus*) is the smallest of the world's otters and is about the size of a domestic cat. Its range stretches from India, through Southeast Asia and up to the Philippines, Taiwan and southern China. This is the otter you are most likely to see – but not in the wild. Like many other otters around the world, it is endangered. The Asian Small-clawed Otter is particularly sociable and lives in groups of up to 12 family members. Perhaps because of its sociability, it adapts well to living in zoos and wildlife parks, and makes a great ambassador for raising awareness about the plight of otters around the world in public displays. So if you see an otter in captivity or at an event, it is more than likely that it is an Asian Small-clawed Otter and not our much bigger native Eurasian Otter.

The Asian Small-clawed Otter has hand-like paws that are not fully webbed, and the short, small claws on its long toes look rather like fingernails. Its paws are highly sensitive and incredibly dexterous, enabling the animal to catch fish between its paws (rather than in its mouth) and winkle out snails from their shells. When resting it hones its skills by playing with sticks and 'juggling' with

Above: Asian Small-clawed Otters are much smaller than their British cousins. Sensitive, partially-webbed paws with 'fingernail' claws are extremely dexterous.

Opposite: Unlike our single-living Eurasian Otters, Asian Small-clawed Otters live in large, family groups. They adapt well to captive breeding programmes, public displays and appearances, helping raise awareness of the decline and conservation of otters worldwide.

Above: A close encounter with a captive Asian Small-clawed Otter can set you dreaming and on a lifelong quest to see otters in the wild.

pebbles, rolling them across its body while lying on its back and relaxing.

The Asian Small-clawed Otter lives in a variety of habitats, from mangrove swamps to rice paddy fields, but is under threat from habitat destruction and pollution. Deforestation, drainage of swampy forests and pesticide pollution from intensive agriculture such as tea and coffee plantations all affect the species, while aquaculture brings it into conflict with fish farmers who cannot afford the fences to keep it out. It is also hunted for its fur and body parts, which are believed in some localities to have medicinal properties.

Fisherman's friend

The Smooth-coated Otter (*Lutrogale perspicillata*) lives alongside the Asian Small-clawed Otter in the wild, and also lives in small family social groups. In southern Bangladesh fishermen have relied on it to help them catch fish, and have bred and trained it to do so for centuries. Three or more animals work together on long leads and harnesses from the river boats to chase fish, prawns and crabs into nets, and are rewarded with titbits of fish. However, for the same reasons why this species is endangered, this centuries-old way of life is in danger, too.

Below: Nimble fingers and teamwork. Bangladeshi fishermen work with their tame, trained otters on the River Ganges, as they have done for generations. The otters work together on leads for a share of the fish.

Everybody say 'aah'

Above: Soft gold: a Sea Otter grooms its luxuriant fur – the densest of any animal – helping insulate it against a life spent in the ocean.

The Sea Otter (*Enhydra lutris*) is undoubtedly the cutest member of the otter family, with the most endearing habits – but of course this cuteness has a practical, evolutionary purpose. The Sea Otter has adapted to living almost entirely at sea, eating, sleeping, hunting, mating and even giving birth in the water. Although our own Eurasian Otter will live and hunt in coastal areas, the Sea Otter's body and way of life have evolved very differently. It can only be found along the coasts of the Pacific Ocean of North America, Alaska and Russia, and it is the only mustelid not to have those all-important smelly, messaging anal scent glands – which would be pointless in a life spent out at sea. Although small for a sea mammal, it is the heaviest member of the mustelid family, at up to 45kg (99lb), more than four times the weight of our Eurasian Otter.

Unlike seals and like all otters, the Sea Otter has no layer of insulating fat (blubber) to keep it warm, so it has evolved the most astonishingly thick fur. With up to 150,000 strands of hair per square centimetre (about a

Rocks in pockets

Otters are inquisitive and intelligent – and the Sea Otter is among the few mammals that have learned to use tools, in this case rocks, to access food. It will search for and pick up a rock from the ocean floor, along with hard-shelled sea species such as clams, mussels or spiky sea urchins, which it will pop into its pockets – deep rolls of skin that form pouches under each foreleg. On surfacing, one animal is taken out of the pocket, and the rock is placed on the Sea Otter's chest as it floats, belly up. Holding the animal between its forepaws, the otter smashes and pounds it against the rock until the shell breaks and it can get at the soft meat inside. The Sea Otter often keeps hold of the rock, popping it back in its pocket to use again, and where it will be handy when it needs it to knock things like abalone molluscs off rocks.

million hairs per square inch) its fur is the densest of any animal on Earth and gives it the appearance of a teddy bear. As in all otters, its fur is so thick that water cannot penetrate through to its skin – even though it spends so much time in the sea.

Having such a luxurious and remarkable pelt, it seems inevitable that the Sea Otter was hunted greedily for its fur. From the mid-18th to the early 20th century, Sea Otter fur became highly fashionable among the wealthy around the world and the most sought-after of all animal fur, to the extent that it was known as 'soft gold'. Commercial hunting killed almost all the Sea Otters. Between 1741 and 1911 the world population fell from approximately 300,000 to just 1,000–2,000 individuals. Numbers are healthier now, but the Sea Otter is still greatly endangered and particularly vulnerable to chemical and oil spills at sea.

The Sea Otter lives and hunts among green, seaweedy kelp forests, and spends much of its resting time floating serenely on its back. It lives in large groups and to avoid drifting away on the tide or becoming separated from its group, it wraps itself up or belts up in ribbons of kelp anchored to the bottom of the ocean. For the same reason it sometimes sleeps holding the paw of another Sea Otter, joining up in big floating rafts of dozens of relaxing Sea Otters.

Below: Sleeping on the surface of the ocean, Sea Otters prevent themselves drifting away from family groups by linking paws, or wrapping themselves in belts of kelp, anchored to the seabed.

Amazonian giants

Above: In a land of giants, these otters share their habitat with Giant Anteaters, Black Caimans, Capybaras, Anacondas and even giant water lilies.

The Giant Otter (*Pteronura brasiliensis*), at some 2m (6½ft) from head to tail, is the largest otter in the world; but with a population of below 5,000, it is also one of the rarest. It lives in family groups along the Amazon River and in the Pantanal Wetlands in South America. The Giant Otter differs from most otters because it is diurnal – out and about in daylight. It is known for its large, slightly flattened, paddle-like tail, and its habit of 'periscoping': raising its long neck up from the water, chin up, on

Right: Each Giant Otter bears unique throat markings. Known as 'River Wolves', they hunt in packs, identifying one another by their cream neck patches by throwing back their heads and rising up from the water.

greeting another one of its kind, or on investigating anything new. Each Giant Otter has individual cream markings on its throat and it is thought this is how members of this species recognise each other.

The Giant Otter sleeps, plays, travels and hunts in social groups. Like the Asian Small-clawed Otter, which does the same, it has developed the noisiest, most sophisticated language of any otter. It barks and growls, wails, screams, whistles and coos, and makes loud, explosive snorts at signs of danger or in warning. It is also known as the river wolf, another canine reference, because it hunts in a pack. This is because it has evolved to tackle animals that are much larger and sometimes more dangerous than itself, such as South American crocodiles, caimans, anacondas (the world's largest snakes) and piranha fish. It is a case of eat or be eaten!

The Giant Otter makes riverside 'campsites' near good feeding spots, flattening and trampling space in the undergrowth, sometimes clearing and collecting branches, then digging burrows and dens within the area. It marks the boundary with communal latrines (toilets), much like Badgers do.

The Giant Otter has the shortest pelt (fur) of all the otters, but it is still dense, velvety and waterproof – and therefore much prized. Despite being protected, it is still illegally killed for its fur. Habitat destruction, including the felling of rainforests, logging, mining, and encroaching agriculture and fishing also threaten the Giant Otter. Although it is highly sensitive to human disturbance on the whole, this big carnivore can sometimes show aggression.

Below: Standing up, the Giant Otter could look many adults in the eye and may show aggression if provoked. Now highly endangered, they are protected by The Karanambu Trust, on the Rupununi River, Guyana.

Sociable otters

The North American Otter (*Lontra canadensis*) and Sub-Saharan African Spotted-necked Otter (*Lutra maculicollis*) are generally solitary, but families of mothers and cubs are often bolstered with siblings from previous litters or even unattached, unrelated youngsters and non-breeding male 'helpers'. Large groups of males can also form to aid hunting in both species and can reach up to 21 individuals. Unlike the Eurasian Otter, they do not defend a territory.

Like the Asian Short-clawed Otter, the African Clawless Otter (*Aonyx capensis*) and much larger Congo Clawless Otter (*A. congicus*) have almost no claws at all. Their toes are also unwebbed and are used, dexterously and sensitively, like fingers.

Below: Most species of otter live alone, but some, such as these Giant Otters, live in groups, making them seem socially more like their badger cousins.

Critically endangered

Sadly, the Marine Otter (*Lontra felina*) found on the Pacific coast of South America is down to fewer than 1,000 adults, while the Neotropical Otter (*L. longicaudis*) and the South American Southern River Otter (*L. provocax*) are also close to extinction. The Hairy-nosed Otter (*Lutra sumatrana*), found in Southeast Asia, was thought to be extinct until recently – but still, just a few survive.

Above: Marine Otters inhabit tidal reaches of rocky seashores along the coasts of Peru, Chile and Argentina. Little is known about them and they are increasingly rare.

Below: The Hairy-nosed Otter hovers on the brink of extinction with just a handful of sightings in nature reserves across Vietnam, Borneo, Thailand and Sumatra.

Eurasian Otters

The Eurasian Otter ranges from Western Europe, east through Asia and on to China and Japan. Its fortunes vary in different countries, but like most other otter species it is under threat and has become extinct in some western countries. In Albania, for example, the Otter is declining but widespread and is still legally hunted for its fur. It is doing well in Greece and thriving in Portugal and the south and west of France. Austrians have given it full protection since 1947 and its population seems to be expanding there, while reintroductions have been made in Holland. In Germany the Otter is highly endangered, and in Italy it is the most endangered animal, with most of the tiny remaining population found in just one 3km (2 mile) stretch of river. It is considered extinct in Luxembourg and Holland, and is missing from the heavily polluted rivers of Belgium. It is a sobering thought that this could have (and almost did) become the fate of the British Otter.

In Britain the Otter is recovering from the brink of extinction. It was relatively common up until the mid-1950s, but then a deadly combination of agricultural pesticide use, water pollution

Above: Sibling companionship for these young Eurasian Otters. Their fortunes worldwide are mixed – and generally declining. But they are staging a comeback in Britain.

and the destruction and tidying up of riverside habitat for agriculture, industry, housing and farming caused its numbers to crash during the 1960s and '70s. It was not given legal protection until 1978. Numbers reached an all-time low in the 1980s, when it was absent from most of England. Small numbers remained in Wales and north and south-west England, but the only thriving populations were in parts of Scotland and

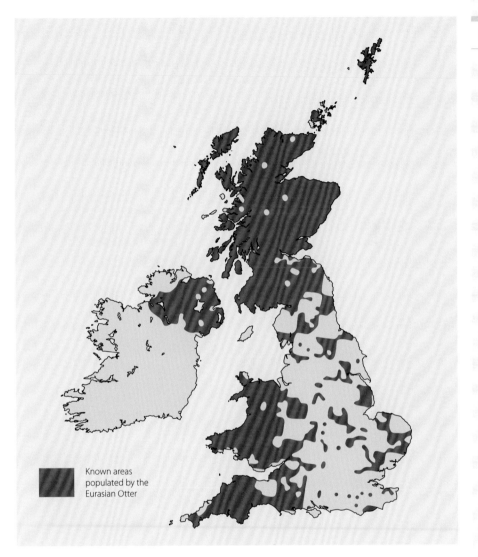

Known areas
populated by the
Eurasian Otter

Ireland. With concerted efforts in conservation and some reintroductions (see pages 72–85), the species has made a gradual but sure comeback and is now present in every county in England, as well as remaining in good numbers in Ireland, Wales and Scotland. It is heartening to think that despite it being an elusive, highly secretive animal, you have a better chance of seeing one now than anyone has had since the 1960s.

Above: The Distribution of the Eurasian Otter in the UK.

Food and Home: a Riverine Life

Many of our towns, cities and villages were built up around water, which we cannot do without. If a river runs through our lives and the places we live, it is quite likely Otters swim through it too.

Otters rely on having a good, healthy habitat – a home that provides all they need to live, breed and thrive, with food, water and somewhere safe and undisturbed to shelter, rest and raise their young. The watery veins, arteries and capillaries of a landscape are at the very core of why Otters look the way they do, how they move and what they eat. Otters have evolved and adapted over thousands of years to live in and alongside water, finding a niche that few other mammals use exclusively, and exploiting it. The fact that Otters are there at all is a good indication that the rivers we live near are healthy for us, too.

A good living

Otters need long, connected stretches of clean, unpolluted water to hunt in. These include rivers, streams, canals and coastline, supplemented with areas that allow them to move about and breed, such as lakes, pools, ditches, reedbeds, fens and marshes.

An area needs to be big in order to hold enough food and shelter to support such a large and energetic animal. An Otter's long, linear habitat will often encompass good-quality areas rich in food and shelter, as well as some patches that are not, which may act just as a means

Above: The river, its tributaries, banks and watershed provide home, food and a means of travel for Otters.

Opposite: Otters are highly adaptable, but they need clean water for a food chain to exist and hiding places to shelter and breed in.

Right: A wet mosaic. Pools and waterside vegetation provide a variety of opportunities in which to find food as well as a place for insects and fish to thrive.

Above: A manmade drainage ditch acts as a connecting corridor for the Otter and a means of travelling from one part of its range to another.

Below: Otter country: the rushing River Elan in the Welsh Cambrian Mountains: full of life and plenty of bankside roots, boulders and crevices for an Otter to hole up in.

of travelling from one part of its range to another. A large range also allows for local events that might disturb and disrupt an Otter temporarily, such as flooding, accidental pollution or disturbance (perhaps that caused by a weekend riverside music festival, or dogs in a favourite dog-walking area), to be avoided and overcome by moving on to a different part of its home range.

A healthy wetland needs plenty of vegetation growing in and around it. Insects will thrive in such a place, as will the fish that eat the insects and the Otter that eats the fish. Waterside vegetation, scrub and old trees also provide vital shelter for an Otter to rest, dry out, hide and shelter in.

What do otters eat?

Otters are piscivorous: roughly 80 per cent of their diet is fresh and fishy, caught live and supplemented by whatever else they find living in their environment. They very rarely eat carrion (dead animals). However, they are always on the lookout for an easy meal – and a chance to expend as little energy as possible. Because of this their diet is both opportunistic and seasonal. They eat a lot of whatever comes their way, or what they find when foraging as they travel up and down their territory.

Otters' preferred food is eels, particularly those up to 50cm (20in) long. Eels are slow-moving and rich in fat, so make a good return for an Otter's efforts. Otters tend to catch smallish, relatively slow-moving, bottom-living fish up to 12cm (5in) long. Their small mouths and sharp teeth can handle these more easily than large and fast-moving fish.

Diving Otters may push out a large fish, such as a chub or perch lurking mid-river in the waterweed, and give chase; or typically, nosing about in reeds or between rocks underwater, they may snaffle up a few bites of 'tiddler' if they come across a shoal of minnows or sticklebacks. Otters dive to the river bottom to hunt

Above: Otters are fish-eaters, but like all good mustelids, they are also opportunists and will eat a lot of whatever comes their way.

Below: Prey varies with location: on the coast, crabs are slow moving and plentiful, but they are a mouthful. It is easier for this Scottish Otter to land its catch and deal with it on the loch shore.

for crayfish, freshwater mussels and fish such as bream, tench and gudgeon. They turn over rocks on riverbeds to snatch up stone loaches, or cruise the bottoms of ponds where toad-faced bullheads sleep during the day. Young, inexperienced fish are taken, but Otters can reach speeds of 12km/h (7½mph) underwater, which sometimes enables them to catch bigger, faster fish such as adult pike, carp, salmon and trout. However, this is rarer and riskier, requiring longer and more energetic chases: an Otter needs to have energy to burn, or to be sure of its chances, in order to pursue larger fish such as these.

Above: Whether in the sea, a river or pond, Otters choose the easy option and forage for smaller, slower, bottom-living fish. High-speed underwater chases expend a lot of energy and are avoided if possible.

Big fish and tricky meals like crayfish and crabs, from which the meat needs to be extricated from shells and claws, are landed and eaten on the bank. They are carried up from the dive in a hug against the chest, but smaller fishy snacks are held between the front paws and eaten whole in the water, fishbones, scales and all. They are tossed to the molars at the back of the mouth on one side or the other, and crunched up by the powerful jaws, with the Otter holding its head in the air.

In spring Otters take advantage of spawning frogs and toads. During the summer months waterbirds feature higher up on the menu. Ducks sitting on eggs or moulting feathers are an easy target. Ducklings and slower-moving, inexperienced young birds may well get snatched from under the water. At these times, the amount of fish in an Otter's diet may be as low as 50 per cent.

Otters' diets vary with where they live, too. Coast-hugging Otters eat small and abundant fish such as rockling, butterfish, pollack, saithe and eelpout, as well as crustaceans like crabs. On the Somerset Levels, many more waterbirds are taken from the dense reedbeds. In salmon rivers, Otters enjoy an autumn bonanza as the fish return to the river to spawn. On Shetland, Otters eat a lot of Rabbits. While they expend some energy chasing them, they have the benefit of being able to hunt on dry land and stay out of the cold water for a while.

Diversifying to survive

Otters are skilled hunters, yet these honed, powerful and streamlined animals will not turn up their whiskery noses at slugs, snails, dragonfly larvae and even the occasional water vole.

On riverside wanders through wet woodlands and other land-based explorations, Otters might find and eat newts or grass snakes. They are capable of surprising turns of speed out of the water (they can outrun a man), enabling them to catch mice and voles.

It may sound like a modern mantra, but Otters have always had to diversify to survive. Living on our islands for thousands of years, they have had plenty of time to learn to do that. A crash in the local population of a favourite fish, for example, or a boom in another, seasonal fluctuations and variations, floods, ice and drought are all things they have learned to deal with very well. Otters, like their environment, are fluid animals, and being able to go with the flow and stream around obstacles when it comes to eating only makes them more resilient.

Above: Occasionally, Otters catch and eat Water Voles. But these two endangered species have co-existed for thousands of years and the number taken is very small.

A word on eels

In the recent past studies of Otter poo (spraints) revealed that Eels made up 80 per cent of the fish eaten by Otters, but eels are now in steep decline. It is estimated that just 5 per cent of the numbers present in the 1980s remain. Eels are enigmatic, mysterious creatures. Hatched from eggs in the Sargasso Sea in the middle of the North Atlantic Ocean, the larvae are swept along by the Gulf Stream until they reach European coasts. Here they develop into small, transparent 'glass eels', and begin to migrate up our rivers and streams where they develop first into elvers, then into adult eels. When fully mature and often many years later, they return to the Sargasso Sea to spawn and probably die.

There are many theories about the deeply worrying decline of eels, as well as concerns about how it is impacting on Otters, given that eels have been such an important part of their

Above: We know what Otters eat and how their diets change from studying the remains found in their 'spraints' or poo.

diet. But the resourceful Otters simply find ways to adapt, expanding their diet. Invasive American signal crayfish, which cause great damage to riverbanks by burrowing and out-competing or eating native species, are booming – and the Otter is quite happy to deal with these easy-to-catch creatures.

Hunting to survive

Above: Water draws heat from the body faster than air, so the longer Otters spend hunting in cold water, the more they need to eat.

Otters are constantly alert for food and new opportunities to find it. As they move around, eating is never far from the agenda – and for good reason. An Otter has a high metabolism: its body converts food into energy quickly, like a fast-running, high-performance engine that needs a constant, good-quality source of fuel.

Because Otters spend a lot of time in a cold, wet environment they must eat a considerable amount to keep their bodies warm, their organs functioning and their energy levels up. They also need plenty of energy to hunt active prey and to cope with lively sections of rivers, tidal currents, swims upstream and getting in and out of the water around weirs, sluices and locks (think how hungry we feel after a swim).

Depending on the water temperature and time of year, Otters need to spend several hours per day hunting, eating around 15 per cent of their own body weight in food each day – more if it is cold or if a bitch Otter has cubs. At about 1–1½kg (2½–3¼lb) of food a day, that is a whole pile of sticklebacks, snails, carp, trout and crayfish *without* their shells.

Chalk streams in southern England, for example, keep an almost constant, relatively warm temperature of 10°C (50°F). Even so, Otters living there need to spend between three and five hours a day hunting, catching around 100g (3½oz) of fish per hour. However, in Scottish coastal waters, where the average winter temperature might be 6°C (11°F) lower, Otters must spend approximately six hours hunting to replenish energy lost while swimming in cold water.

Otters' lives therefore involve a careful and precarious balance. The longer they stay in cold water, the more they need to eat – but of course if they cannot find enough food, they need to stay in the water longer. Life is tough for this high-performance top predator.

Fast food and takeaways

Some meals appear to be there for the taking, even when they are associated with man. A small garden pond full of goldfish or slow, expensive koi carp can provide an easy meal if it is on an Otter's beat. Although Otters prefer smaller fish not favoured by anglers, fish farms, trout hatcheries or a lot of big fish in a lake stocked for fishing are sometimes taken advantage of.

Above: Although Otters handle smaller fish better, salmon-rearing pens, like these on the Isle of Mull may provide easy pickings for a determined Otter.

Competitors for food

Otters have no real competitors for food, except perhaps the North American Mink. Non-native and invasive, mink moved into the ecological gap the Otter left as its numbers declined. But as Otters have been reclaiming their habitat, Mink seem to have declined. In some areas Otters co-exist with Mink, but there is also evidence that Otters fight and even kill the smaller Mink. There are also indications that Mink have diversified their own diet away from fish and have moved into habitats away from water. The displacement of mink by Otters is good news for the highly endangered Water Vole. It does not often feature on an Otter's menu and the two species have co-existed happily for thousands of years. Mink, on the other hand, are small enough to squeeze into the bankside burrows of water voles and decimate populations.

Territory: holding the line

Above: An Otter's long, narrow territory can only be defended by constant vigilance and travel.

Depending on how healthy a river and its environs are, and how many other Otters are in the vicinity, a male Otter's long, linear territory can be 20–40km (12–25 miles) or more in length. It often overlaps with a female Otter's territory, which is typically half that length.

Otters are fiercely territorial, but it is impossible to defend such a vast area all of the time and an Otter will always be several kilometres away from most of its territory. Otters may also spend a week or more in just one area if the living is good at a particular time.

In order to defend and hold a territory, Otters live nomadic lifestyles within the boundaries of their range, constantly patrolling up and down their patch, and usually covering 11km (7 miles) of river each night. With the addition of frequent naps, roughly three-quarters of an Otter's time is spent asleep. However, Otters cannot afford the time or energy to keep going 'home' to sleep – and risk losing one end of their territory and all it contains to another Otter. So Otters require a series of up to 30 quiet, sheltered, undisturbed places within their home range in which they can clean and dry their fur, lie up and rest, have lengthy periods of sleep, and conserve energy levels and warmth.

Below: Most of the Otters' time is spent napping to restore energy levels and warmth and any safe patch will do, from a favourite hiding place to a more temporary spot.

Holts and hovers

A holt is a more permanent and often underground den that may have been used by generations of Otters. It is often formed in tunnels and hollows caused by water erosion in the roots of bankside trees, the large root cavities of fallen or old trees, or perhaps an old Fox earth or enlarged Rabbit burrow a couple of bounds from the water. The Otter prefers a snug, tight, secluded space and can squeeze its long, slim shape through a 15cm (6in) hole – it rarely needs to dig. Some holts have more than one entrance and may be accessed by water. Otters will also happily climb up into the low crown of a coppiced or split Crack Willow.

Other resting places, known as hovers, are more like temporary campsites for this water gypsy; somewhere dry, and protected from sight and the elements. Flood debris, bankside vegetation, overgrown heaps of tangled branches or hollows under Hawthorn or Blackthorn scrub all provide protected areas for Otters to snooze the day away, as well as part of the night, as do crevices, nooks and crannies in rocks, or simply an overhang of a bank. Bedding is not required unless a bitch Otter is breeding, and cold concrete does not put them off. Otters readily make use of man-made constructions such as drains, culverts, mills and ledges under bridges. Man-made holts are particularly appreciated on waterways where there is a lack of cover – kilometres of river or canals full of fish are of no use to Otters unless there is shelter, too.

So called 'couches' are even more transient lying-up places where an Otter may just curl up above ground against a tree trunk in a bay of roots, or on a flattened 'nest' of reeds. A patch of bracken may provide a place where an Otter can roll, squeeze and wring the water out of its fur, and snooze in a warm, sunny spot.

Below: A snug, secure, underground holt, hidden by bracken and marked with an accumulated pile of droppings (a spraint heap) may have been used by generations of Otters.

Agility and Grace: a Life in Water

Many writers and observers of Otters talk about how the animals appear so totally absorbed in their watery element and seem completely immersed in just 'being Otters'. Perhaps this 'sense of an Otter' tells us more about the human spirit and our fascination with an air-breathing mammal that is so at home in wild water, much as birds and their ability to fly does.

The Otter does not just move and hunt in water, it revels in it. For us, the Otter embodies the essence of freedom, joy and independence within an element that we all crave a little. The Otter is at one with the medium it has evolved to live in. It is perfectly attuned to its environment, passing through it with dolphin-like fluidity, slipping from the land into the water as if being poured in. Indeed, in much earlier times there was some confusion over whether the Otter was in fact a fish and not a mammal at all.

Above: An Otter has a sleek body covered in fur ideal for slipping in and out of the water with consummate ease.

Adaptation to swimming

The body of the Otter is beautifully streamlined. Its wide, flat head is wedge-shaped to offer little resistance to the water and is smaller than its neck. The narrow but powerful shoulders follow on with the slim ripple of the Otter's long, strong back, big rump and muscular tail that tapers to a point.

When an Otter cruises along it sits very low in the water with most of its body submerged, for the least detection. Its head is sunk in the water below its cheeks, but its nose, eyes and ears (aligned, head-on, in a 'V' shape from its nose) are positioned high on its head, so that these highly tuned sensory organs sit in a horizontal line just above the level of the water like the Plimsoll line on a ship.

Opposite: The Otter swims so low in the water that only the top of its head and the powerful curve of rump and rudder are visible, giving it the profile of the Loch Ness Monster breaking the surface.

Above: With its highly–sensitive ears, eyes and nose aligned just above the surface, the Otter is aware of everything around it and able to react quickly.

An Otter's shoulders and most of its back dip beneath the surface, but the top of the high rump is often visible, as is the tail, giving the impression of three slight bumps in the water – it looks hump-backed, like a low-profile Loch Ness Monster. The progress of a cruising Otter appears smooth, stealthy and effortless. It barely disturbs the water, leaving only a widening 'V'-shaped wake behind it.

Under the surface the toes of the Otter's big paws (on average 5cm/2in across) are splayed wide to make them more efficient and allow the webs between them to open out. The Otter 'doggy-paddles' much more smoothly than any dog ever could, its big, webbed hind feet powering it along.

With its nose, eyes and ears in the air, an Otter can cruise quietly along while remaining alert to danger – at a sign of it the Otter can simply sink, and is ever ready to dive.

A diving expert

Like an amphibious craft, an Otter adapts quickly when diving. It can close its nostrils and fold its tiny ears shut, making them watertight. It then swims down by kicking with all four feet. For more power and speed the Otter 'porpoises', leaping partway out of the water like a porpoise or dolphin, and diving. Tucking its forepaws back against its body and aligning its hind feet against its tail, it undulates its long, flexible spine so that its whole body and tail move in a serpentine wave. It steers by using its long hind feet and the powerful muscle of its rudder-like tail. In this torpedo-like shape, an Otter can reach speeds upwards of 12km/h (7½mph).

Most dives that an Otter makes last for less than a minute, but it can stay completely submerged for four minutes at a time to forage along the bottom of a river or shallow lake, or slip beneath the water to escape a perceived threat. It can also swim 400m (1,312ft) on a single breath – there one moment, gone the next. The Otter has evolved large lungs in order to take such deep, lasting breaths, but also has an advanced dive reflex that allows it to slow its heartbeat underwater and use less oxygen.

Below: The streamlined Otter can swim underwater for 400m (1,312ft) on a single breath, although most dives to forage in the river or on the seabed last less than a minute.

AGILITY AND GRACE: A LIFE IN WATER

An Otter's pelt

Above: Seaweed provides a great rolling place for this Otter to squeeze excess water out of its coat to start the drying process.

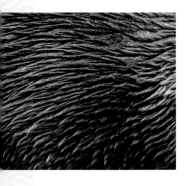

Above: A double-layered coat keeps the warm air in. Once out of water, guard hairs join together forming arrows that direct water away from the Otter's body.

Even in the best of summers the Otter's wet environment is a cold one. Water conducts heat away from the body 25 times faster than air does. Having long fur might keep an Otter warm on land, but long fur would cause too much drag in the water and take a dangerously long time to dry. A blubbery layer of fat under the skin such as a seal's would be impossible and impractical to maintain on such a fit, muscular and active animal as an Otter. The Otter has therefore evolved a sleek, thermally insulated, waterproof fur coat, or 'pelt'.

The Otter's coat consists of two incredibly dense layers of fur. The undercoat is packed so thickly with short hairs that they form a kind of soft, felted wool. Over the top of this lies a smooth layer of longer guard hairs. Combined, this double-layer coat contains a phenomenal 70,000 hairs per square centimetre. Compare that with a Husky dog bred for snowy climates, with its great thick coat: it has just 600 hairs per square centimetre!

Thermal insulation reduces heat loss between two extremes of temperature – the Otter's warm body and the cold water. Air warmed by the Otter's body heat is trapped against its skin in the felty undercoat, and prevented from escaping by the close layer of long hair, effectively sealing in the air. The long hairs are also lightly coated with natural

water-repellent oils in the Otter's skin. Water never even reaches the skin, making it totally waterproof.

On diving, the pressure of the water squeezes the Otter's coat flat and surplus air escapes, giving the animal a fine, fizzy aura of silvery mercurial bubbles. A chain of bubbles bursting on the surface is often the only indication that an Otter is swimming past.

The sleekness of the Otter's close, smooth, velvety pelt eliminates most of the 'drag' that swimming causes – water just slips over its lightly oiled and glossy coat.

As soon as an Otter is on land, the coat immediately opens up because of the air still trapped in it. Helped along with a good shake, the guard hairs form spiky arrows to direct water away from the Otter's body and allow more air into the coat to speed up the drying process. The Otter often has a good roll to squeeze and wring the water out of its coat, followed by another shake.

Grooming then becomes particularly important. The coat must be cleaned, and the Otter's natural oils replenished and spread back over the pelt to re-proof it.

If an Otter has been swimming in the sea it must rinse its coat in fresh water afterwards to prevent the sticky salt from seawater from compromising its high-performance protective coating.

Above: Grooming is a serious business, cleaning and spreading natural oils back over the coat reproofs it. Coastal Otters bathe regularly in freshwater to remove sticky saltwater deposits.

Above: A lightly oiled, glossy pelt makes for a warm, waterproofed, sleek, and efficient Otter. An unkempt Otter would quickly become dangerously chilled.

Underwater vision

Above: Otters can adjust the focal length of their eyes to focus underwater, helping them to hunt in clear water.

The Otter is an animal of two worlds – one in and one out of the water. It uses its senses differently in each medium, adapting them to whichever one it is in.

Much like a Badger, an Otter cannot see much more than shape and movement on land, but unlike a Badger it does have good underwater vision. Adapted for air, the human eye can only focus underwater behind goggles or a mask that keep a layer of air between the eye and the water. Specially developed muscles behind the Otter's eye adjust the curvature – and therefore the focal length of the Otter's eye lens – to the different refractions of light in water. The Otter's eyes become rounder, more convex and more fish-like, enabling it to see its prey in clear or well-lit water.

Like many other mammals, birds and other animals, the Otter has a nictitating membrane, a third eyelid that slides sideways across the eye like a barn door. It protects the eye underwater and is completely transparent, allowing the Otter to keep its eyes wide open when fishing.

However, even sparkling clean water is a poor conductor of light compared with air. Since the Otter hunts mainly at night and rivers and lakes are often murky, a sense of touch (and possibly smell) is of far more use to a hunting animal. Yet being able to see well underwater is still a useful tool in the Otter's sensory toolbox.

Feeling in the dark

The Otter has a wonderful array of strong, long whiskers, called vibrissae, all around its muzzle. Bristling like a Victorian gentleman's moustache, they act like antennae in the dark. Although the whiskers themselves have no feeling, the hair follicles they grow out of are highly sensitised with nerve endings.

A little wider than an Otter's body, vibrissae help the animal to feel its way in the dark, alerting it to obstacles. When foraging along a riverbed an Otter can tilt its whiskers down and forwards for precision location of any prey. Vibrissae are also highly tuned motion sensors that can detect the presence of a fish from a mere wobble in the water. Acting like trip wires, the whiskers subtly vibrate in the Otter's muzzle so that even the size and species of a passing fish can be determined by its wake as it swims past.

Below: Bristling with information and highly sensitive nerve endings, the Otter's impressive set of whiskers can detect the subtlest of movements in water, acting as trip wires in the dark for otherwise invisible, passing fish.

A tickle on the nerve endings from a couple of whiskers can cause an Otter to whip round in an agile 180-degree turn and grab an otherwise invisible fish.

Blowing bubbles, sniffing water

Above: An Otter uses all its senses to check for danger before breaking cover and it is likely all these senses come into play beneath the surface, too.

It is hard to tell just how important an Otter's sense of hearing is underwater, but what about its sense of smell? New research indicates that the Otter might use this sense – which is acute above ground – underwater too.

Otters have been filmed locating food by blowing a bubble of air out of the nose towards objects in the water and sniffing it straight back in. Wildlife film-maker and seasoned Otter watcher Charlie Hamilton James has witnessed an Otter locating a fresh dead fish at night in this way, when neither sight nor a sense of touch were of any use. It is also suspected that Otters dive with a bubble of air under their top lip, in case it is needed for the same purpose. Water shrews and North American Star-Nosed Moles are known to use their sense of smell in this manner, so it is possible that Otters do so too.

Efficient teeth

Above: With sharp, directionally opposed, prong-like canines, even the slipperiest of eels is secured. An Otter's teeth slice, grind and crush through flesh, shell and bone.

The Otter has a small mouth and very sharp teeth. Its front canines are long, pointed and set at an angle leaning away from each other, which makes them excellent prong-like tools for gripping slippery eels and other fish, and smooth, hard shells.

Once prey has been caught, it is thrown to the back of an Otter's mouth to be crunched up by crushing molars. These wide carnassial teeth have slightly rounded surfaces and blade-sharp edges that work to crush and scissor their prey. From the side, an Otter's teeth can look like a set of fierce pinking shears; they are efficient enough to crush a crab carapace or bite through a fish's backbone.

The Otter out of water

Otters often travel across dry land between rivers and wetlands, crossing roads and diverting around sluices, weirs or lock gates, for example. Their gait is quite unlike that of any other animal. On land their powerful rump is higher than the head. Otters' backs are too long and their big hind feet too large (compared with the front feet) to allow the animal to trot like a dog, with its feet moving in opposite diagonal pairs. Otters therefore bound on land and can get up a good gallop. They will also stand upright, balancing on the tail, to look around and check for danger, ever ready to slip back into the water and melt into the current again.

Above: The elongated, hump-backed, bounding run of an Otter on land is unmistakable, if surprising. Here, sharp claws grip the frozen lake to build up speed, but Otters often choose to slide on ice and snow.

Family and Play

This sleek, sinuous, full-bodied animal, chasing and chuckling, gripping things with a gurgle and leaving them with a laugh, to fling itself on fresh playmates that shook themselves free, and were caught and held again.

Kenneth Grahame, *The Wind in the Willows*, 1908

The above extract from *The Wind in the Willows* describes Mole's first experience of a river – but it could just as easily be describing the Otter we might get to glimpse. However, Otters are something of a paradox. They are playful and sociable, yet aggressive and solitary; highly mobile but fiercely territorial; secretive but boldly 'signposting' their presence to other Otters, and inadvertently to us, too. Bitch Otters are attentive, exhaustive mothers, while dog Otters simply pass on their genes.

Meeting up

Eurasian Otters live a largely solitary life and when they do meet, their behaviour seems unpredictable to us. They sometimes greet each other with enthusiasm and affection, whickering and playing before moving on; at other times they react to one another decisively with violence. It is thought that family members and previous partners are generally welcomed, unless a female has cubs or there is competition for food or space. Unknown Otters prospecting for new territory or just passing through an occupied beat are confronted immediately.

Above: Adult Otters are solitary animals. Dog Otters leave their territory to find a mate, but soon return, leaving the female to raise the cubs alone.

Opposite: Mother Otters form strong, close bonds with their cubs and the family stays together for more than a year. They are affectionate and playful, with the mother often joining in or even instigating play.

When a dog Otter picks up the scent of a female in season, he leaves his territory to find her. A bitch Otter can breed in her second year, and both male and female Otters have several partners in their lifetimes. They can breed at any time of the year, although spring and late autumn are favoured.

Otters acquaint themselves with each other in what looks very much like a game. They play and show off (there is tail waving from the male), chase each other along the banks, roll together, and swim, splash and dive. Mating takes place in the water.

A dog Otter stays with a bitch for just a short time – a few days or a week at the most – before returning to patrol his own territory. He takes no part in raising his cubs at all.

Single mothers

Female Otters have a nine-week gestation period and during this time they are particularly secretive, finding a quiet, discreet holt away from the main river in which to give birth and raise cubs. It is often up a side stream or a little way from water in a place that is not prone to flooding. The maternal holt is very snug and often consists of just a slightly enlarged rabbit hole. This is the only time that Otters use bedding, and the mother gathers moss and

Below: The entrance to a Rabbit burrow, or the discreet, snug, nursery holt of an Otter and her cubs? The spraint heap outside gives it away.

grass in long rolls to make a nest in the holt. In expansive reedbeds a nest is sometimes made with a domed lid or thatched roof constructed from reeds and grasses.

Above: Cubs are not taken out of the holt until they are at least three months old. They are nervous, inexperienced and stay close to their mother.

Between two and five cubs are born, although infant mortality is high, with often only one or two cubs making it to adulthood. The tiny cubs, weighing just 40g (1½oz), are born with their eyes closed and their bodies covered in a thin layer of grey fur. They are helpless for the first five to six weeks. At six to seven weeks, their eyes open and they start to take solid food.

Otter cubs grow and develop slowly; they stay in the holt for up to three months, only venturing out to toilet. Although this may seem like a long time to stay 'indoors', their mustelid cousins, Badger cubs, make their first forays out of the underground sett at around the same stage. Otter cubs are weaned from their mother's milk at around 14 weeks.

Mother Otters work incredibly hard to raise the cubs alone, hunting for eight hours or more a day to feed themselves and their cubs; consequently, they are more likely to be seen out during the day than at other times. Mothers may move their cubs one by one to different holts if there is a threat of disturbance, discovery or flooding.

Swimming lessons

Above: A mother teaches her cubs to hunt – but also to swim. She chooses quiet lakes and calm water to begin with.

Surprisingly, Otter cubs are not born swimmers. They do not go into the water until they are around four months old. A bitch Otter's range usually includes quiet, calm reaches of water such as lakes, pools, canals and tributary streams, where there is less energy in the currents than there is in faster-flowing water bodies.

Cubs have to be taught to swim and it is not easy at first. They are fluffy and buoyant with 'puppy fat', so are initially good at bobbing in the water, but have trouble diving beneath it. They are also instinctively cautious, even nervous of the water at first. Sometimes reluctant Otters have to be dragged or dropped into the water. Being born wary of such a changeable element makes good evolutionary sense. Otters are still semi-aquatic, air-breathing land mammals that have adapted to hunt in water. Early over-confidence could easily result in drowning. The cubs stay with their mother for between a year and 18 months. In that time cubs get to know the ways of the river and tides in all their moods – spates, floods, droughts and currents – and the skills to negotiate and exploit them, as well as learning how to hunt by copying their mother.

Anxious cubs can sometimes be seen swimming so close to their mother that they touch, so that a long line of Otters forms, like one big animal; or, if they are pestering her for food or just playing, they will swim side by side as a 'raft' of Otters.

Schooling youngsters to hunt

Otters are curious, playful animals and when they are cubs, play is a way of learning, practising and honing skills that will be used for survival as independent adults. Cubs spend a lot of time playing with their siblings, as well as (to a lesser extent) their mother. They playfight, roll and tumble on land and, more often, chase, dive and splash together in the water.

Just as curiosity lends itself to investigative behaviour and success in finding new sources of food or shelter, to a degree adult play is useful for maintaining skills and quick reactions, as well as keeping animals fit and supple. Friendly adult Otters have been observed romping and swimming together in an affectionate manner when they meet, weaving and twisting around one another as they swim.

Below: Play fighting. Youngsters hone their skill and agility, strength and fitness through tussles with siblings and their mother. On meeting old acquaintances, adults play too, keeping their own reactions sharp and themselves in shape.

Otters create and frequently use mud slides to get into the water. Sliding down a bank uses marginally less energy and perhaps makes less noise than running down it, but Otters have occasionally been observed doing it more than once, surely just for the fun of it! Similarly, in snow or on ice Otters sometimes choose to slide.

Some of the more sociable species of otter are regularly seen playing with objects in captivity. Although these otters are well fed and do not have to work hard to find their food (and therefore have time to play), wild Eurasian Otters have also been spotted rolling and catching stones, shells or small sticks between their paws and mouths. This all helps to maintain a level of dexterity, coordination and alertness, and can only help in catching a slippery, shadowy, fast-moving meal.

Above: Whilst hunting enough food to stay alive is no game, play helps to sharpen responses, reinforce muscle memory and quicken reactions.

Whistling over the water

Male and female Otters call to each other with high-pitched whistles. The calls carry above the noise of a river, and are similar to those used by waterbirds like Kingfishers, Grey Wagtails and Dippers. Most of the vocalisations made by Otters are contact calls between cubs and their mother: a high-pitched whistle from her and bird-like chirrups, cheeps and whistles from them. Cubs also make twittering noises during play that escalate into cat-like noises when they (or adults) fight seriously. A loud 'hah!' is produced as an alarm call, and is often followed by a chittering threat sound.

Striking out

The cubs usually leave their mother and start to disperse from 14 months of age. They are now as big as her (or bigger, if they are male), but are reluctant to leave. Just as she had to push them into the water initially, she has to push them out of her territory.

Sometimes the cubs are chased away by their mother or depart when a dog Otter comes to find her – but she may also run away from them to find a new mate, or even a new territory, leaving her old one for the cubs to sort out.

Below: Leaving home to establish a new territory is a tough and dangerous time for a young Otter. It makes good sense to keep a low profile.

A corridor of bullies

In a linear habitat it is easy to imagine that all would be well if the cubs moved into the territories 'next door' and all the Otters just shunted up one. However, good territories have to be fought over, and the gene pool would be very limited if Otters of the same family lived too close to each other – so families need to spread out. Juvenile Otters probably have to run the gauntlet, passing through the established territories of several aggressive neighbours to get to a new one.

Sadly, many young Otters are killed at this stage. They may get into fights with other Otters and have to travel long distances, crossing roads between watersheds in order to find unoccupied and decent feeding grounds. They may make do with an inferior patch for a while, pioneer new land and water, or remain transient until they can settle. Otter life is harsh.

Above: Passing through other Otters' territories is unavoidable and risky in such a narrow habitat. It may take months for a newly independent Otter to find a home.

Scented messages

Otter poo is so much more than a bodily function to get rid of waste. Known as a spraint, it is an entire and complex messaging service, a note left stuck to the fridge door, a business card, a boundary marker, a love letter.

Otter poo is quite small for such a big animal (about the size of a little finger), but with a diet of many small, carnivorous meals, there is a lot of waste to excrete regularly: fish bones and scales, shells, and the bones and inedible parts of small mammals and birds. Some droppings are deposited in the water, but all adult Otters make a special effort to leave some on dry land.

Parts of the long territory of an Otter can go unvisited for quite some time, so that kilometres of it may at times be left physically undefended. By depositing spraints in prominent places along a river, an Otter informs all other Otters that this patch is taken and *will be defended*, preventing a deadly fight. Spraints are also used to issue challenges within an occupied territory, and by bitch Otters to indicate that they are ready to breed, inviting dog Otters into their own home range.

Below: A fresh, tarry spraint, lubricated with jelly-like mucus that eases the passage of such food waste, also contains a detailed, social messaging service for the Otter community to ignore at their peril.

Spraints include more than 100 different scent components and, along with scent glands beneath the base of the tail and in the feet, contain 432 different chemicals. Urine and a jelly-like substance produced from the lining of the stomach (which helps lubricate and ease the passage of all the sharp, scratchy waste) are also important in scent marking.

We can only guess at the information that is shared and passed on among the animals in this way, but it is thought to detail age, sex and the general health and condition of each animal. It is also assumed that individual Otters can be identified by their unique spraints.

Otters go to a lot of trouble and take great care about where and how they leave their spraints. To a solitary, secretive animal, scent is everything – and scent is mostly lost in water.

For maximum effect, spraints are left in prominent places and any areas where ownership may be ambiguous, such as the confluence of two rivers, a junction of ditches or lakes away from the main stretch of river. On each beat there will be favourite sprainting places – some of which will have been used by generations of Otters. The spraints must act as 'scent bonfires' or beacons, posted somewhere obvious and

Above: An Otter leaves spraint in a prominent place where it will be noticed and there can be no confusion over whose patch this is.

Below: A spraint heap used by successive owners of this patch. In the absence of boulders or large tree roots, Otters use raised grass humps, ant or mole hills or make grass and earth 'plinths'.

raised just high enough to highlight their presence, and for the scent to drift like a smoke signal across the sensitive nose of a passing or swimming Otter. Exposed rocks in a river, low, overhanging branches and prominent roots, bankside tumps, hummocks, anthills and molehills all make ideal sprainting places.

Because spraints may need to last a week or more, they are often left above tidal reaches, or protected from the elements by a bridge or natural or man-made overhang. In winter they are positioned on higher ground to avoid rising water levels.

If a plinth or platform cannot be found in the right place, Otters build sand castlings or haycocks out of grass to spraint on. Otter paths are also marked with little grass sentries, so an intruder can be in no doubt as to who owns the territory and all its rights of passage.

Otters spend a lot of time checking and 'reading' messages left by others, as well as carefully renewing their own. Transient juveniles and mother Otters and their cubs – which do not want to be detected for fear of aggression – hide their spraints by depositing them in the water.

Below: A fresh, tarry spraint looks very different from an older, pale, crumbly one; but the sweet, fishy scent identifies it, lingering long after its owner has gone.

Fighting among Otters

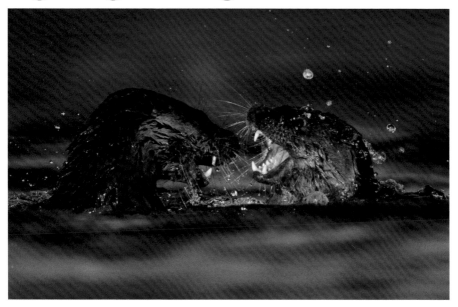

Very many fights between Otters are avoided through the sprainting 'messaging service', but if an Otter enters another's territory having ignored the 'keep out' signs, a fight inevitably follows.

Fights are serious and fierce. Both sexes fight over territory and a female also fights to protect her cubs. Dog Otters can be a threat to cubs and may even kill those that are not their own.

Fights need to be decisive: an Otter cannot stick around in one part of its range to keep 'warning' another Otter off, and injuries are often horrific. The animals try to get underneath each other to attack the soft belly and reproductive organs of their opponent. Wounds can become lethally infected and most road deaths are the result of one Otter fleeing another, or are of Otters that are slowed down by an injury picked up in a fight.

There is evidence that fights between Otters are becoming more frequent – this is thought to be down to competition among increasing numbers of Otters and the availability of good habitat.

Above: Not all fights can be avoided. There is little hesitation and they are usually serious, taking place on land, in water and often both.

Below: Fights over territory need to be decisive and Otters can horribly injure each other. Dog Otters will even fight females that are smaller than them. Conflicts can lead to road accidents and wounds can become infected.

Almost Extinct: the Otter's Past

Throughout history Otters were presumed 'guilty' of taking 'our' food. They were the sworn enemy of monks with stewponds full of fish, large country estates with fishing rights, and fishermen, rather than an indicator of the health and vitality of the river and all of its food chains, including ours. Over the centuries Otters were largely treated as many top predators were: as a general pest, to be hunted down and killed as 'vermin' and – as an occasional source of fur – trapped and turned into clothing, gloves and hats.

Yet all this time Otters were still widespread throughout Britain. They were hated by some, cherished and observed by a few, and accepted by most as part of our wildlife and the river landscape that sometimes ran through human settlements, largely unregarded and unconsidered.

After the 1950s, however, in a very short space of time all the Otters in most of the rivers in England and Wales, and to a lesser extent Scotland and Ireland, mysteriously disappeared. We came close to losing our Otters completely.

Opposite: Until relatively recently Otters were considered pests by many, who felt they deserved to be hunted, trapped and shot for stealing fish from the lakes and rivers of country estates, such as Sandringham.

Right: Coming close to losing our Otters completely, we now consider them a sign of a healthy riverine environment and all its food chains, including ours.

Hunting with hounds

Above: Otterhounds have a kind nature, but are bred to hunt, with a loud baying 'voice' and rugged stamina that carries them all day over land and in water.

In 1157, King Henry II of England appointed a King's Otterer and gave him both a large manor in Aylesbury, Buckinghamshire, and the responsibility of organising regular Otter hunts with hounds. A specialist breed of dog, the Otterhound, was developed in a tradition of hunting Otters that would continue into the 20th century.

Henry VIII's Preservation of Grain Act, 1532, encouraged the killing of animals listed as vermin by paying a reward bounty for each animal killed. During the years of Henry's reign, a human population boom coincided with a run of bad weather and poor harvests, resulting in food shortages and diseases such as plague and a deadly flu. The 'vermin list' was wide ranging and based on huge inaccuracies and assumptions. Hedgehogs were on the list for drinking milk from sleeping cows and for rolling around in orchards to 'steal' fruit away on their spikes. However, Otters were not legally considered vermin until 1566, when Elizabeth I revised and lengthened the list. Our observations of wildlife and its prospects did not improve: bounties were now paid not only for Otters, but also for the heads of scavenging, carrion-eating Red Kites – surely no threat to our food supplies? Yet to many, particularly the rural poor, payment for killing mammals and birds on the vermin list meant a meal for the family.

Right: Unthinkable now, a Victorian illustration from 1860 records the death of a hunted Otter. Spears and tridents were used to kill Otters into the early 1900s.

In 1653 Izaak Walton published his famous book, *The Compleat Angler*, announcing: 'I am, Sir, brother of the Angle, and therefore an enemy of the Otter; for I hate them perfectly, because they love fish so well.' Walton was respected and his book, still in print today, was widely read. It confirmed to many that hunting Otters was both necessary and to be encouraged.

Above: Crowds gather in Trafalgar Square in 1978, the year Otter hunting was banned, to protest against seal hunting in Canada. The Otters' dramatic decline prompted hunts to cease voluntarily, two years earlier.

From the 19th through to the mid-20th century, Otter hunting became popular among some rural riverside communities. Packs of hounds were hunted on foot, and the hunt became the subject of Victorian paintings and was considered something of a 'spectator sport'. Riverside picnics were even organised around a day's hunting, taking place in the summer months, unlike winter foxhunting. Sometimes, if an animal was perceived to have given 'good sport', it was allowed to go.

However, by the early 20th century opposition to hunting Otters was growing and becoming more organised. Otter hunts were declared particularly cruel because there was no 'closed season' as there was for other quarry such as Foxes, hares and deer, when hunting stopped during the breeding season. Mothers and cubs were sometimes killed. One part of the hunt involved followers standing next to each other to form a human wall, or 'stickle', using sticks and poles to turn Otters back towards the hounds.

At the height of its popularity, around 450 Otters a year were killed by hunting, but the intention was not to kill all the animals in an area. Otters were an accepted part of the riverine environment. Unsurprisingly, huntsmen were the first to notice a sudden lack of Otters in rivers. They tried to bring their discovery to national attention and ceased hunting Otters entirely in 1976. Some packs switched to hunting mink until that was banned in 2005.

The vanishing Otter

The disappearance of such an elusive animal from our waterways was difficult to prove or quantify, so the first national Otter survey (England) was commissioned in 1977. It was undertaken by Libby Lenton, a teacher and naturalist who travelled the country in her camper van and walked the riverbanks, looking for signs of Otters. Even before the results were collated and published, it became obvious that the situation was dire, and Otters became protected, by law, in 1978. The results of the survey were shocking. Libby had searched over 1,264km (785 miles) of riverbank without any evidence of Otters. In fact, in nearly 95 per cent of the sites where Otters had been present, there were now none. Scotland, parts of Wales and Ireland still had Otters, but in England just 170 scattered places showed any evidence of the animals, mostly along the Welsh Borders and in south-west England. Within the space of 20 years barely 6 per cent of English rivers had Otters living in them. What had happened?

Below: Within 20 years our rivers emptied of Otters, 95 per cent of their former range had no Otters in them at all. Wildlife Trusts such as this one in Staffordshire have led the way in conserving habitats to increase Otter numbers.

Population crashes

It had taken a dangerous amount of time to establish that Otters were in trouble, but while their numbers continued to freefall it took more time to work out why, and even more time to do anything about it.

Initially an increasing human population and the rise of house building, industry, water extraction and leisure (such as fishing and boating) were variously blamed. Riverbanks were being built on and rivers were straightened and canalised.

These factors were certainly having a negative impact, but they did not explain the dramatic, rapid disappearance of Otters from rivers that were free from all these characteristics. At first, clues from a seemingly unrelated source were missed – and they were quite literally falling out of the sky.

For some time since the 1950s, farmers and gamekeepers had been concerned by the unexplained deaths of Woodpigeons and game birds such as Partridges and Pheasants. The RSPB began investigating a suspected disease. At the same time numbers of Peregrine Falcons, and indeed other birds of prey, had been falling dramatically. In 1956 the situation was confirmed with a huge population crash in Peregrines. But no one yet could either explain the deaths or link them.

Above: Otters were being squeezed out of their habitat, but the connection was not made between their decline and what was seeping into our rivers, or pouring from the outfall pipes.

A chemical romance

Since the mid-1940s what seemed like miracle chemicals had been gaining in popularity in farming and industry and had begun to be used widely. Dichlorodiphenyltrichloroethane (DDT) had been developed to prevent malaria and typhus in soldiers fighting in the Second World War, and after the war it continued to be used as an insecticide, along with other organochlorine pesticides. Deliberately poisonous to insect pests, weeds and fungal infestations, these pesticides were used as seed and bulb dressings, sprays on all manner of crops from cereals to apples, and 'dips' to rinse sheep of blowfly larvae infestations and scab, a serious skin infection caused by mites. However, these pesticides, which often had no taste, smell or colour, did not just target the insects or weeds that were considered pests; they were quickly soaked up by the soil and washed away, giving the rivers a huge dose of poison each time it rained.

Farming was becoming more industrial and farm slurry (liquid manure, which used to be spread on the land) was often left to wash into rivers, while chemical fertilisers were used to encourage the growth of crops instead. By-products and chemicals from factories, including polychlorinated biphenyls (PCBs) used in paints, plastics, fire retardants and coolants for engines, were simply rinsed into the waterways as if they were one great drain and not a living organism.

Above: New 'wonder' chemicals from farming and factories were poorly understood and increasing in popularity, but the run-off and by-products of them were poisoning the rivers.

The effects of the chemicals were poorly considered and understood, and at first no one thought to blame pollution or pesticides for the loss of the Otters, or to make the link between the deaths of game birds and the decline of the Peregrine.

The rivers and all the life in them died. Just two spoonfuls of some of the chemicals could kill several kilometres of invertebrate life in a river, and they were being used in unnecessarily high doses. Otters became extinct in the Netherlands, Luxembourg and Belgium, and came horrifyingly close to extinction in France, Germany and Italy. However, the picture was starting to come together. In 1962 American author and ecologist Rachel Carson published her famous book *Silent Spring*, highlighting the impact pesticides were having on our wildlife and human health.

Birds such as Woodpigeons and Pheasants (and other animals) were eating chemically treated grain and the poison was being 'stored' in their considerable fat reserves, killing them only when these were depleted during cold spells in winter. It was hard to detect.

However, the food chain of the Peregrine is much shorter than that of the Otter and it was discovered that the chemicals did not disappear quickly, but instead accumulated as poison up the food chain. A top, athletic predator such as a Peregrine has little or no fat reserves and the toxins built up in the reproductive organs instead. A report in 1968 found that Peregrine eggs were either infertile or that their shells were too fragile.

The pesticides were to blame, but it took another decade until anyone realised that the same thing was happening to our rivers and the Otters.

Above: Peregrine Falcons were the first obvious victims of chemical poisoning, with the toxin load accumulating up the food chain, yet it took another decade before this was attributed to the Otter's decline.

Eels, Otters' favourite food, are very high in fat and consequently take on and store a huge amount of toxins in their flesh, including heavy metals such as lead. Like falcons, Otters have small fat reserves and the pesticides 'bioaccumulate' (intensify up the chain) and build up in their reproductive organs. Their immune systems are also weakened, making them vulnerable to diseases, including new ones brought in by escaped mink. Many Otters suffered blindness, and cubs ended up drinking concentrated poisons in their mothers' milk.

A pattern emerges

It took the UK more than ten years longer to ban DDT than America, Hungary, Norway, Sweden and Germany. Other organochlorines (Dieldrin, Heptachlor and Aldrin) were already in use as replacements for DDT, but these proved just as deadly and poison continued to leach into the rivers. In turn, these were replaced by all-new organophosphates that broke down more quickly in the environment, but were more toxic. Synthetic pyrethroids like cypermethrin replaced the old sheep dip in the 1990s after concerns over the health of sheep farmers, but these were found to be 1,000 times more toxic to invertebrates. James Williams, author, naturalist and Otter hunter turned Otter conservationist, stated that: 'In areas where enhanced synthetic pyrethroids have been used as

Right: Run-off from animals 'dipped' in chemicals to kill mites, ticks and lice, also killed the insect life for miles downstream, and seriously affected the health of sheep farmers, too.

a sheep dip, invertebrates can be completely absent. One damp sheep can sterilise miles of a stream completely.' Synthetic pyrethroids were banned in 2006.

Wildlife was an unintentional victim of the new wonder chemicals, but so were we. As well as losing the wildlife we were really beginning to value, the chemicals were found to be dangerously harmful to human health.

For 60 or more years, from the mid-1940s to 2006, we had been killing our rivers and everything that lived in them.

Our rivers, lakes and coastal areas are now healthier and cleaner than they have been in decades. Even the River Thames, in London, which was declared 'biologically dead' in 1957, is the cleanest it has been in 150 years and the wildlife is coming back.

Localised pollution, deliberate or accidental, continues to threaten wildlife. Many chemicals persist for decades in the silt of rivers and the flesh of long-lived eels. We are continuing to develop chemicals, pesticides and even medicines that find their way into our rivers as low-level pollution. New neonicotinoid pesticides are thought to be partially responsible for the huge and worrying decline of Honey Bees.

Above: As we continue to develop new chemicals, we must be aware of unwanted side-effects. Bees are in steep decline and nobody really knows why.

Things we have learned

- Nothing lives in isolation. If one species is in trouble, or a link is broken in the food chain, it will impact on other species.
- If a top predator is in trouble, things are very serious indeed.
- We are part of nature and at the top of a food chain, too.
- We should take notes. By recording what we see and submitting our results to different surveys and wildlife organisations, we give the experts an early warning when things are going wrong, allowing us all to do something about it faster. We can also celebrate the successes.

Threats and Recovery

The recovery of our Otters from the very brink of extinction is one of the great conservation success stories of the 21st century. It is a source of wonder and celebration that we are still willing and able to accommodate such a large and specialised predator on our crowded islands. Things could have been very different. However, the story is far from over.

The recovery of Otters was, to begin with, painfully slow and is not by any means complete. The 21st-century Otter, and those who want to protect it, still faces many obstacles, challenges, new threats and unknowns. With the lessons of the recent past still ringing in our ears, continuing research, vigilance, education and appreciation are vital.

In the Otter's V-shaped wake is the story of how we got it very wrong, but also how we are beginning to get it right. Otters are the physical embodiment of second chances.

Opposite: Once we realised we did not want to lose the Otter, huge efforts and greater understanding have begun to bring it back from near extinction.

Below: A bow-wave of hope in its v-shaped wake, an Otter reclaims its former territory.

What happened next

Since 1978 Otters have been legally protected under both the (amended) Wildlife and Countryside Act 1981 and the Conservation of Habitats and Species Regulations 2010. Their breeding and resting places are also fully protected, to the extent that it is an offence to even disturb them.

Perhaps unsurprisingly for an animal that lives at a low density, the recovery of Otters happened slowly. Many factors were to blame for their decline that would take time to put right, and of course for a long time, as soon as harmful pesticides were banned, new damaging ones were taking their place.

The second national Otter survey carried out between 1984 and 1986 showed a small 10 per cent increase in Otters – things were starting to go in the right direction.

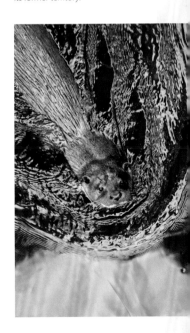

Reintroductions

In the early 1980s it was hoped that the pockets of Otters in Wales and the healthier populations in Scotland and Ireland would naturally recover, but it looked as though the animal would be completely lost from most of England. In 1971 Philip and Jeanne Wayre established the Otter Trust on their farm on the banks of the River Waveney on the Norfolk–Suffolk border, where they cared for rescued, injured or orphaned Otters. From these Otters they developed a captive-breeding programme at three sanctuaries at home, and others in Durham and Cornwall. Between 1983 and 1999, along with the government's Nature Conservancy Council (which became English Nature and is now Natural England), they reintroduced these animals into some of the rivers. A total of 117 Otters were released, mainly in the intensely arable area of East Anglia, where pesticides had all but wiped out

Below: A concerted effort between wildlife trusts and charities, government bodies and volunteers, reinstated the habitat that supported the spread and recovery of Otters and small-scale reintroductions.

the Otters. Six animals were released in Dorset, Wiltshire and Wessex, and the last release was of 17 Otters on the Upper Thames catchment in 1999.

The project came under some criticism because it was thought that the rivers had not recovered sufficiently to support the animals, but many survived to breed and boost the remaining native Otters, particularly in East Anglia.

In the meantime, the Vincent Wildlife Trust created 'Otter Havens'. Working alongside other wildlife organisations such as The Wildlife Trusts and the RSPB, as well as water-supply companies and the Environment Agency, they protected key Otter habitats and replanted riverside vegetation to allow a recovery and expansion. They were additionally instrumental in carrying out further Otter surveys across the UK. Injured and orphaned Otters were also rescued, rehabilitated and re-released by the trust along the River Derwent in Yorkshire.

By the early 1990s, the big clean-up of rivers, the protection and restoration of habitat, and the hours spent by volunteers building Otter holts where there was little bankside cover or vegetation, had paid off. The Otters were coming back.

By the third national Otter survey (1991–4), it had become apparent that Otters were beginning to make a good recovery by themselves, spreading out from strongholds in Scotland, Wales and south-west England. Reintroductions were no longer deemed necessary (and ceased in 1999), and conservation efforts were concentrated on continued habitat management and improvement.

An otter in every county

With great fanfare and celebration, by 2011 Otters were present in every county in England. Kent had been the last Otter-less county and now had a pair.

The return of the Otters was the ultimate stamp of approval, and their spraints and pawprints in the river silt were the 'seal' on the health of our rivers and wetlands. The broken links in the food chain, from vegetation to insects, fish and Otters, were being mended.

Above: The Otters' return was slow but by 2011, 33 years after the alarming results of the first National Otter Survey, Otters were recorded in every county in England.

Urban otters

Excitingly and increasingly, Otters are reappearing in rivers where they run through our big cities and towns – and where they have not been seen since the Industrial Revolution in the early 1800s. Their appearance, often in broad daylight and in the centres of a lengthening list of big cities such as Birmingham, Dundee, Newcastle upon Tyne, Reading, Cardiff, Leeds, Bristol, Edinburgh and Manchester, seems to challenge what we 'know' about Otters – that they are shy, secretive, nocturnal animals, highly susceptible to disturbance.

It may be that Otters are moving into urban areas because other, quieter stretches of the river are already occupied. Perhaps they are becoming more tolerant of people; they may have no collective memory of hunting and persecution. Some individuals may have reassessed their relationship with us in the search for food and shelter, and discovered that we are not a threat with our riverside football games, picnics, shopping trolleys, bikes and prams. They may have recognised that our soft and pampered pet dogs are not the tenacious hunting dogs of the past – and are generally no match for the surprise of a territorial Otter and the threat of its lutrine teeth.

Similarly, Otters may be out during the day because they have to spend more time hunting down food if it is not plentiful, or because they accept our noise and bustle as non-threatening. Perhaps, hidden in plain sight, they simply feel safer.

This much we do know: that Otters are extremely adaptable, intelligent and individual animals, and our urban rivers and canals are now cleaner and healthy enough to support them. There is a chance that while out shopping, among the traffic, the sales bargains and the pushchairs, you might glance down a river when crossing a bridge and be rewarded with an Otter sighting.

Below: Some modern Otters are more tolerant of people and are increasingly seen in daylight in urban areas where they are easily overlooked among the bustle.

A cautious glee

The presence of Otters in every county is of course wonderful news – but there is still a long way to go before they become widespread again and it is debatable whether their recovery might ever be considered 'complete'. Otters are all but impossible to count in the wild because of their normally nocturnal, highly mobile and secretive ways, differing lengths of territory and the difficulty in telling them apart. National Otter surveys are done by dividing the country into 10 sq km (4 sq mile) grids. Lengths of river bank where Otters were previously evident, measuring 600m (1,968ft), are then checked for spraints or pawprints. This provides an indication of how well populated an area is, but cannot tell us how many Otters there actually are. They may be roaming more widely, perhaps across several grids, in search of food and shelter. Between 2008 and 2011 the survey showed an astonishing 44 per cent increase in Otter spraints in the Ribble Valley, Lancashire, which the media, suitably excited, interpreted as a 44 per cent increase in Otters – but it is impossible for Otters to breed that fast. Otters are thought to spraint more frequently when there is less food about. A lot of what we 'know' about them is still informed guesswork.

Above: Otters readily make use of manmade structures such as drains, culverts and bridge supports. They seem to have no qualms about sleeping on concrete.

A low population

Above: A single Otter needs a large territory to provide it with enough good hunting. This loch may support just one Otter, or a single mother and her cubs.

The strongest populations remain in Wales, Scotland, parts of Devon and Cornwall, and East Anglia. Some of the rivers in these areas are thought to be at their 'carrying capacity'; that is, they are fully occupied by Otters. A single Otter takes up a lot of space and Otters are 'self-regulating': they do not keep squeezing up until there is no more room, but limit their own numbers. In areas that are almost 'full' there may be more competition for space and food and an increase in fighting, but weaker or younger Otters just move further away, or settle for lower quality territories.

Wild Otters do not have long lifespans, living on average for just four years (compared with 10 or more in captivity). Because cubs often stay with their mother for over a year, bitch Otters breed every other year, not until they are two, and may only have two small litters in their lifetimes. The population of this top predator remains necessarily low. Our rivers are unlikely to ever be packed with Otters.

Fisheries and garden ponds

In some circumstances, the return of the Otter has proved problematic and has even been unwelcome. Otters were still considered pests that could be legally culled within living memory. In the intervening years of living without Otters, much has changed. Many people have forgotten, or not known, what it is like to share a habitat with a wild, native natural predator, which has been around far longer than us. Fish farms developed during a time when Otters were rare or absent from rivers, and the need to protect fish from such a predator was not considered. But here was a confined, captive source of readily available, easily accessible food – a handy Otter takeaway.

Many Otters ignore fisheries, or stop by them once in a while as part of a patrol of their territory, while other areas are heavily predated, particularly if they have been found by a mother with cubs, causing losses of thousands of pounds to a business.

'Put-and-take' fisheries, where large stocks of fish are released into rivers to be caught later, and 'specimen' fisheries, where there is a high density of very large, often single species of fish in small pools and lakes, are an easy target for Otters. Carp and catch-and-release fisheries, where fish are fed and grow old and fat, are particularly vulnerable. Although non-native, carp have become naturalised since being introduced in the 1300s by monks for food. 'Trophy' carp are very long-lived and can grow to an enormous size. They become sluggish in winter and provide an easy meal for Otters, which catch them by the head. Similarly, Otters may discover koi carp or goldfish in garden and ornamental ponds. These often very expensive fish are frequently large and have slower reactions than wild fish; and they are all confined in very small spaces.

Below: The confined space of small ponds full of slow moving fish, such as these ornamental Koi carp, provide easy pickings for a hard-pressed Otter.

A big fish might be found on the bank with just a few chunks taken out (where an Otter has only eaten 'the best bits' like the heart and liver) and cubs, like small children, are messy eaters, leaving lots of fishy leftovers. Otters rarely come back for fish they have killed and half-eaten, because by the time they are hungry again they are already several kilometres away. All this can seem very wasteful to an angler.

Fair game?

Angling is very popular in the UK and contributes billions of pounds to our economy: a single fish can be worth hundreds of pounds. However, few of us, including few anglers, seek to put Otters back on the vermin list. They are one of our favourite animals, and much time, dedication, effort and money have been invested in their recovery. Otters are good for tourism, but more importantly than perhaps anything else, their presence tells us that we are beginning to fish in healthy, sustainable waters.

Below: A Grey Heron stalks the shallows. Otters are not the only predators of fish, but an expanse of quality habitat should mean plenty for everyone.

Can we blame Otters? They are not the only predators of fish – herons, cormorants, signal crayfish and Mink can cause significant losses. And there is still a lot of work to be done on improving habitat and invertebrate life for fish in rivers so that all life can be supported. For the sake of the future conservation of Otters, we must seek to understand the conflicts and sympathise with problems, putting ourselves in the chest waders of others so that we can listen, engage and debate, then ultimately work out a way in which we can all enjoy the water together.

A Fox in the water

Learning to live with Otters – or fencing them out – may be the only option. Fencing fisheries can prove both difficult and costly. Otters are resourceful and can dig, jump, climb and use overhanging branches to scale inadequate fences, and fencing rarely looks attractive, particularly in the case of ornamental ponds. But much as any poultry farmer – or anyone else who keeps chickens or ducks – knows, good fences are the only way to keep out Foxes. Foxes are not legally protected, of course, and can be shot. But Otters are more territorial, so removing an individual would create a vacuum that would soon be filled by another Otter.

Above: Only a good, strong fence will keep a hungry, determined Otter out of a fishery or a garden pond. We should expect nothing less of this tenacious, top predator.

The ghost of Isaak Walton and his Otter-hating mantra of 1653 may still haunt a few anglers, but the more enlightened know that if the quality of the river habitat, and the fly and fish life, are good enough, then Otters and anglers can co-exist.

Debunking myths

Articles in the recent press have shown evidence of the kind of backlash that often comes after a success story. The sensationalism and inaccuracies contained in these articles help neither anglers nor Otters, detracting from the real issue of promoting and protecting healthy river life. Otters may be described as cute, but they have never been 'cuddly'. Some Otter facts:

- Otters are indicators of the overall health of our rivers and wetlands.
- Otters will rid most areas of Mink, which cause much damage to fish and other wildlife.
- Otters eat non-native signal crayfish that

eat small fish, fish fry and eggs, as well as damaging native wildlife and the riverbank environment.

- Otters benefit fisheries by taking sick or injured fish, improving the health of a fishery.
- Reports of population explosions are greatly exaggerated: 'lots' of Otters seen together will be a mother and cubs.
- Only a small number of captive-bred Otters were released, and no Otter has been released since 1999.
- Otters do not tend to kill more than they can eat in an instinct to save food as a Fox does – Otters do not cache their food.

Current threats

Above: Naturally inquisitive, Otters explore eel Fyke nets, crayfish traps, and lobster pots and creels, sometimes becoming stuck and drowning. Many fishing devices are adapted to prevent this happening.

Drowning, traffic and the dangers to eel populations are all serious threats to Otters today.

In the past, many Otters were inadvertently caught and drowned in Fyke nets, a series of netted funnels used to catch Eels. A contribution the Vincent Wildlife Trust made to Otter conservation was to develop Otter guards for the nets, make it illegal not to use them and freely distribute them to eel fishers.

Small young Otters (and indeed water voles) sometimes get caught in mink traps, despite Otter guards being fitted to prevent their entry. Mink are caught live, then humanely despatched at the earliest opportunity, giving the trapper a chance to release other animals caught by mistake. However, Otters frequently drown after entering illegal signal crayfish traps and getting stuck. A licence is needed from the Environment Agency to trap the crayfish, and an Otter guard must be used.

Road-traffic collisions are the biggest killers of Otters, and many Otter habitats are dissected by roads. Animals are often forced onto roads to travel around weirs, sluice gates, flood defences or other man-made obstacles, and when river levels are high or 'in spate' they may not be able to swim upstream or under bridges. Cubs may cross roads to avoid trespassing into the territories of others. As Otters spread out and recolonise the country, many are killed by cars; it is thought that in some counties road deaths are keeping up with the production of cubs. Where possible and in known problem areas, ramps up weirs, ledges in culverts, underpasses, tunnels and fencing can all help.

The 95 per cent decline in populations of the eel, until recently the Otter's preferred food, forces Otters to find an alternative food source. No other fish is as rich in energy as the eel, and eels are slow and easy to catch. Their decline is bound to put extra pressure on Otters, even though they are resourceful animals.

Below: Roads bring inevitable tragedies. The Environment Agency and The Highways Agency (which identified 400 potentially dangerous road crossings in South-west England alone) are working to create safe alternatives.

Below: Eels are nutritious and easily caught, but this former staple is in trouble and Otters must search elsewhere for food.

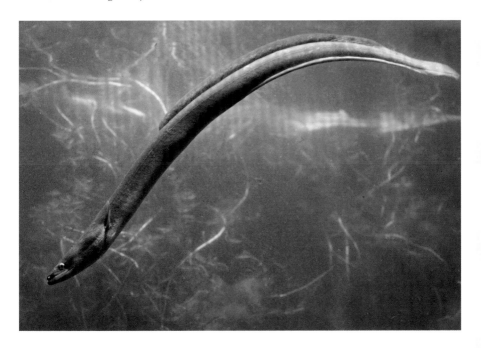

Above: When river levels are raised, such as in a flood, Otters often have no option but to avoid the water involving risky road crossings.

The Cardiff University Otter Project is a national scheme established in 1992 that collects the bodies of Otters found dead in England and Wales for post-mortem examination, from which much is being discovered. Otters are hard to study and do not tolerate collars that might hold tracking devices, so dead bodies and spraints provide vital scientific information.

Managed in collaboration with the Environment Agency, a number of research projects are carried out on the Otters, including monitoring toxins in the Otters' body tissues and research into a recently discovered

1948	DDT begins to be used as an agricultural pesticide.
Mid-1950s–60s	Some Otter hunts and fishermen notice a decline in the numbers of Otters found and seen. The first Otter hunt, in the Wye Valley, closes in 1957.
1956	Big crash in Peregrine Falcon numbers.
1960s–70s	Otter hunts note an absence of Otters in some rivers and raise concerns. Many hunts disband.
1962	Rachel Carson's *Silent Spring* is published.
1968	DDT linked to decline in Peregrine Falcons (specifically as a result of eggshell thinning and infertility).
1973	DDT banned in America.
1976	All remaining Otter hunts take the joint decision to close.
1977	First national Otter survey reveals that just 6 per cent of former Otter territories show any evidence of Otters.
1977	The Vincent Wildlife Trust initiates Otter Haven projects.
1978	Otters are legally protected and Otter hunting is banned.
1983–99	The Otter Trust reintroduces 117 captive-bred Otters, mainly in East Anglia.
1984	DDT banned in the UK.
1984–6	Second national Otter survey shows 10 per cent increase in Otters.
1989	Dieldrin banned in the UK.
1991–4	Third national Otter survey shows 22 per cent increase in Otters.
1992	Cardiff University Otter Project starts: national research based on spraint analysis and post-mortems on Otters, among other projects.
1994	Government's UK Biodiversity Action Plan aims to restore breeding Otters to every watercourse and coastal area where they had previously been recorded, by 2010.
2000–2002	Fourth national Otter survey shows 36 per cent increase in Otters.
2009–10	Fifth national Otter survey shows 59 per cent increase in Otters.
2011	Otters recorded in every county in England.

THREATS AND RECOVERY

parasite, the bile fluke, as well as the prevalence of kidney stones in Otters. Researchers have also discovered traces of hormone-disrupting chemicals that are widely used in common painkillers, such as Diclofenac and Ibuprofen, inside the hairs of Otters. Flushed through the sewers and into the rivers, they are thought to be responsible for reproductive problems now occurring in the animals, including a shrinkage of the baculum, the male Otter's penis bone. It bears repeating: research into this living barometer of our rivers is essential to monitoring our health, too.

Otter Spotting

You can put in hours of searching for the merest hint of otter, and more waiting for it to turn up; but get so much as a glimpse and you realise, you would do it all again.

It is a truth universally acknowledged among Otter watchers that Otters see people more than people see Otters. They are few and far between, and given the lengths of their territories, their erratic sleeping and hunting habits, and their near-silent ability to melt into the water and dive and swim right past you, seeing one is never going to be easy. Yet these magical, tricksy animals are an enigma: you may come across one being watched by a crowd of people (who have not put in the time and effort you have) on a Saturday morning in town. Otters have irregular habits, but may be reliably seen for a week or two in the same spot. They are nocturnal, yet sometimes diurnal; highly secretive, yet advertise exactly where they have been and where they go so that their presence cannot be missed – by you (if you know how) or another Otter. Knowing for certain that they have been around is a bit like finding the hoofprint of a unicorn. They are elusive, but not impossible to see, and the good news is that you have more chance of spotting one now than you would have had in the last 50 years or so.

Below: The Otter is an enigmatic, difficult animal to see, which makes it all the more magical when you do see one.

Opposite: Look, then look again, an animal that seems to take on the colour of the earth, the water and the woodwork can be hard to spot.

How to see an Otter

Above: Welcome to Otter time, where naps are taken at will and movements follow a pattern known only to the Otter.

Otters follow a rhythm all of their own. Their appearance is hard to predict and varies with the place and individual. Their sense of time is different from ours and they seem to have an innate restlessness. They take naps at will, and due to fluctuations in food sources, tides and currents, the movements of fish shoals, and patterns in weather and seasons – coupled with an environment that can alter dramatically within hours or change slowly over weeks – sightings are difficult to predict. They may frequent what is currently a good feeding place every day for three weeks, but then not visit it again for a year. Some places, such as the Somerset Levels and the coasts of western Scotland, can be more reliable than others. This is all part of the allure, challenge and magic of Otter watching. Sometimes the alchemy convenes and you find yourself in the right place at the right time: immersed in Otter time.

Otter watching tips

- Dawn and early mornings, or just before dusk, are the best times to go looking.
- Do not take your dog – Otters are not keen on dogs.
- Do not wear perfume or aftershave, or smell of your tea or breakfast. Otters can smell chicken curry or bacon rolls in your clothes a long way off.
- Wear clothes in dark, earthy colours that do not rustle or flap about. Make sure you are warm enough when keeping still and are protected from biting midges if they are about.
- Take binoculars, a torch at dusk and a small general field guide. You will want to identify plants, birds, prints and droppings around you while you wait.
- Avoid chatting, and agree some sign language with your friends.
- Walk into the wind. If you walk in the direction in which the wind is blowing, an Otter will scent you.
- Tread quietly and slowly, rolling your foot from heel to toe along its outside edge. Pick up your feet so you do not swoosh through the grass.
- Keep a low profile. Avoid being silhouetted against anything, or letting your moving shadow fall on the water. Keep your hands still or in your pockets.
- Wait where you can see Otter paths, or entrances and exits to the water, especially if you notice that they are damp or wet in dry weather. But do not sit too close.
- Let your nose run rather than sniff or reach for the white flag of a tissue – no one is going to notice. Sucking sweets (without crinkly wrappers) can help suppress a cough.
- Learn to fish. A lot of anglers see Otters, and Otters need anglers that like them.
- Bear in mind that it is illegal to disturb Otters, so if you think you know where a holt is, just watch from a safe distance.
- Above all, be careful around water and tell someone where you are going. Curiosity and keenness can easily result in you slipping into the water. A stout stick can help in ground that has a tendency to sink.

Below: Sometimes, the alchemy of weather, time and tide, a shift in the river flow or the movement of fish convenes, and your longed-for Otter materialises.

Listening and watching for Otters

It is often difficult to hear sounds or pinpoint them when around running water or rustling reeds. Cupping your ears helps. Listen for the vocalisations of Otters talking to one another – high-pitched birdlike whistles, twitterings and cat-like noises might give them away. Pay attention to rustling and splashes, and tune in to the wild world around you. The noise of Blackbirds or ducks sounding the alarm, or Moorhens suddenly exploding up from the river, could all be directed at an otherwise stealthy Otter.

Be alert to the fact that you may glimpse just part of an Otter, or even just the movement it is making. A 'V' shape in the water could be the wake of a swimming, mud-coloured Otter that has become part of the current; a chain of bubbles popping on the surface may be evidence that an Otter has seen you, dived and gone. You may just glimpse the strange sight of a disappearing tail, like the wave of a whale's fluke, as the animal dives. Three separate humps in the water may morph into one Otter – or even three in the shadows along the opposite bank.

Below: A ripple on the tide or a ribbon of seaweed shifts into a subtle curve of fur, or the low, flat head of an Otter.

Do not be fooled

Many people mistake mink for Otters. There are a lot more Mink about and they are more visible. They are, however, very different. Otters are much larger, sleeker and distinctly dog-like, while Mink are ferret-like. Otters have broad, flattened heads and long, tapering tails, and swim low in the water. Mink have pointy, 'pencil-sharpened' faces and fluffy tails. They look furry, buoyant and more laboured in the water than Otters.

Below: Beware, your heart might skip a beat for a mink. They swim higher in the water, are fluffier and have pointed, ferret-like faces.

Where to find Otters

Visiting a wetland wildlife reserve that Otters are known to frequent is a good way to start a search for them, and an RSPB reserve is perfect. There may be wardens and volunteers with local knowledge you can plunder, as well as the possibility of hides, webcams and log books recording the latest sightings. It is also worth checking websites before you go.

Sadly, many of our rivers are inaccessible and under private ownership, but there is nothing better than finding evidence of Otters or even seeing them on your nearest local waterway. Friendly anglers might be approachable – ask them if they have seen any, but be prepared for an anti-Otter stance and a chance for you to do some Otter PR.

Bear in mind that when you have almost given up on ever seeing an Otter, one might just appear. Tell everyone you know; they will be as green as duckweed with envy.

Fieldcraft and tracking

Above all, do not give up. All the time you will be accumulating knowledge about Otters and immersing yourself in their world, discovering plants, birds, insects and other creatures that share it. Tracking such magical, mysterious creatures as Otters is nothing short of an adventure, involving scrambling under bridges, sliding down banks and extreme stealth with a quiet snack. Honing your fieldcraft skills and sensory powers of observation (eyes, ears, nose and touch) will lead you to that elusive half-mythical sixth sense, too: a hunch, a feeling that something is there. But half the excitement and thrill of Otter watching is finding your own evidence that Otters have been there. It may be a fresh spraint under a road bridge, or half a petalled pawprint rayed like the sun behind a cloud as it enters the water. On a dry, dusky evening you may get to lay your hand on a wet patch in the grass where an Otter has wrung the water from its pelt, and where the grass is still lifting from the press of its body.

Below: Prints made by the hind- and forepaws of an Otter. Such fresh, clear tracks are rare and even then, the fifth toe or the full pad does not always imprint.

Pawprints

A good Otter pawprint, known as a 'seal', clearly shows five teardrop-shaped toes, curved around the large pad like a sunrise. The pads are large, about 5cm (2in) across, and round or long in the hind feet; they will spread out larger in very soft mud, and look slightly lopsided. The Otter's smallest toe often does not make a mark. A nice, clear silty print may show claw marks and even webs, or the long drag of a tail. The prints of a mother and her smaller cubs might be found together, and those of a dog Otter will be larger. Prints can overlap or be imprinted over and muddled by other animals.

Spot the difference

- **Dog** Only four toes, symmetrical and in hugely variable sizes.

- **Cat** Smaller and rounder prints. Fifth toe does not show. No claws showing.

- **Fox** Four toes. Pawprints narrower and neater than a dog's, and often in a long, straight line like a running stitch.

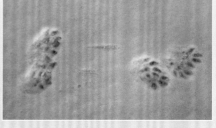

- **Mink** Similar prints to an Otter's, but smaller (about 3cm (1¼in) long). More pointed toes to go with that face.

- **Badger** Similar in size to Otter tracks. All five toes point forwards, and look pigeon-toed. Long digging claws often visible.

Spraints

One of the best things about the way in which Otters communicate, by sprainting regularly in prominent places, is that it gives us indisputable evidence that they are around. Another is the smell. Thankfully for an animal whose poo gives us so much information, it smells nice, or if that is not your cup of (jasmine) tea, at least it is inoffensive.

Below: More shell and bone fragment than anything else, a distinctive Otter spraint can be dissected like an owl pellet.

The smell of a spraint is its key characteristic: no other poo smells like it. It is sweet, fishy and often likened to new-mown hay or jasmine tea or – by me – to ivy flowers. So if you are in doubt about who did it, get down low and sniff.

Otter spraints are only about the size of a little finger. The shrapnel of a variety of fish bones and scales, eel jaws, frog and small mammal bones, feathers, beetle wing cases and pieces of shell are visible, all held together by a sticky black, tarry marmalade.

On sprainting sites you may just find a tiny blob of jelly or, if you are extremely lucky, the milky mucus of a cub spraint (though a cub's presence is rarely advertised).

Below: Get down and think like an Otter. Where would you stick your post-it notes? Somewhere they will last and will not be missed.

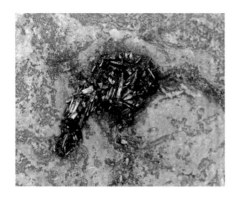

Spraint is often placed where its smelly chemical message is out of the rain and other situations that would destroy it, so it (and its smell) lasts for a long time. Older spraints dry out, becoming oily green, then grey, white and crumbly like ashes.

In the search for poo it helps to try and think like an Otter. Spraints are left at any point where a territory may be in doubt, or at ways in, out and around water: anywhere, in fact, that you might leave a poster advertising a fête for Otters. Look carefully at the largest rocks in a river, at clumps of Greater Tussock-sedge, fallen tree trunks and exposed tree roots, and pay particular attention in places where ditches or side tributaries join a main river, at junctions, weirs and on concrete ledges under bridges and culverts. Also check entry and exit points to rivers (marked by slides or tunnels in the bankside vegetation), and Otter paths over old anthills and molehills, or grassy hummocks that have been utilised by insects. Where there are rivers, shores, reedbeds or lakes without prominent features, Otters scrape up sand and stones into castlings or cairns to spraint upon, or build small haycocks or tumbleweeds out of grass – which then dries out and becomes more conspicuous. A series of these, like little unmanned sentries, may mark a path across a watershed.

Where an ideal sprainting place has been used by generations of Otters, it is known as an 'Otter seat'. These places may be easily identified by either a whiteness on the bank, where accumulations of ammonia have killed moss or lichen, or they can be spotted by a patch of raised, uncharacteristically lush, dark green grass where nitrogen from the spraint heap has fertilised the grass.

Above: An Otter 'reads' a message in a sprainting place. The best way for us to identify Otter poo is also to sniff it.

To be able to distinguish Otter scat from that of mink, note that fur, feathers, bones and fish scales are visible in mink scat. They twist and taper, rather like Fox poo, but are rather smaller, about 1cm (½in) wide. They are dark brown or black, shiny when fresh and often have an unpleasant smell. Older scats lose their scent and are mouldy, not crumbly like those of Otters.

Waterbird poo is sometimes gritty and greenish, smelling of silty river mud.

Chutes and tunnels

Otter paths can be identified because they begin and end in water. Otters usually take the shortest route and run around an obstacle such as a sluice or weir, or cut a corner of a meander or an oxbow in a river. Their paths are narrow and well worn, having sometimes been used by generations of Otters, and are often marked by sprainting places.

Entry and exit points from water will be marked by low, Otter-sized tunnels in the grass. Where a bank is steep, entry points may be worn smooth to form a 'slide' or chute. Sliding into the water saves energy and allows Otters to enter the water more quietly than if they galloped in with a splash.

Left: This is an Otter-sized tunnel through the rushes. Look for regularly used entry and exit points into water and shortcuts around obstacles from an Otter's viewpoint. Paths always begin and end in water.

Advanced Otter watching

If you are lucky enough to have access to a river and you know Otters are about, there are ways to capture the evidence.

- Make a sand or silt trap to capture pawprints. Use a solid wooden board or a tray, spread it with damp sand or silt, and smooth and level it off. Place it securely (you do not want to put the animal off) where you suspect there is an Otter path and check it for prints regularly. Make plaster casts of the pawprints if you are lucky enough to find any.

- Set up a trail camera that will be triggered to take a photograph when something moves past it. Set it near (but not too close) to an Otter path and check it regularly.

- Build an Otter loo. On the shore of a river, build an ideal sprainting place that territorial Otters will not be able to resist – a pile of stones or bricks, or a large rock, might make the perfect loo with a view (for you).

The remains of meals

A large discarded fish carcass left on a bank with the energy-rich vital organs bitten out is often a sign of an Otter meal. The head of a large fish may bear witness to how it was caught, as will teeth marks about 2.5cm (1in) apart in its face. The fish carcass may now have been abandoned by the Otter, but it will not be wasted. Herons, Foxes and Badgers will often claim such carcasses. Crayfish claws and crunched up bits of shell left on the bank are another sign of Otter meals.

Cubs are particularly wasteful eaters during the time they are learning their craft and stealing food from one another. Messy remains and bits of fish can be signs of cub meals.

Otters also push up turf. It is thought that they do this to find worms or larvae just under the soil like their mustelid badger cousins do. Rather than making the deep, snout-sized holes and diggings characteristic of Badgers, Otters roll the turf back neatly. Whatever they need seems to lie at a relatively shallow level – it is possible that they are searching for the taproots of grasses, containing minerals missing from a fishy diet.

Below: Given the nature of their food, Otters are messy eaters and you may come across the leftovers. Smaller catches are eaten in the water, but this large crab will have had to be landed.

Finding the grisly remains of half toads, with their loose spotted skin lying where their legs once were – or even half toads turned inside out – is very lucky indeed. You can be certain Otters have been feeding there. Otters are one of the few creatures that will risk tackling the nasty-tasting skins of toads. Large paratoid glands in the skin behind the eyes of toads contain powerful toxins, and Otters are the only animals that have learned to skin toads in order to eat them. First they bite the hips, pulling off the back legs and eating the muscle before peeling back the skin to eat the insides. They will often treat frogs in the same way, but whether this is because they cannot tell amphibians apart or are just being cautious, it is hard to tell.

An Otter's environment

Above: A rich wildlife habitat means plenty to see and discover. Do not miss the opportunity to immerse yourself in Otter Country; you will be rewarded even if you don't spot what you have come for.

The most wonderful thing about tracking and watching Otters is that they are an inseparable part of the environment that they have adapted to and evolved to live in, and what an environment it is. We humans like being near water. It evokes strong emotions in us and produces different watery responses: being near water can be soothing, relaxing and peaceful, or enchanting and exciting. It provides a chance to forget about one world and enter another.

However, Otters share their environment with other intriguing and wonderful creatures that you cannot ignore, and which you are far more likely to see. They will give you clues as to the quality of the place and what else is about.

Learning your fieldcraft and being observant will result in your getting so much more out of visiting these wet, wild places. People need water. Wildlife needs water. You may be in for a few surprises. And getting close to the other wildlife in these places brings you ever closer to Otters.

Above: Find out beforehand what is likely to be about; it helps with identification and it's exciting to be able to name something like this Scarlet Tiger Moth.

Before you start a search

Have a flick through a guide book or go online to see what else might be about. You will know what to look for and it may help you identify what you see. Remember to take into account the seasons, the type of water (for example a fast upland river, lowland chalk stream or gravel-pit lake) and the habitat surrounding it.

Above: Look for an electrifying flash of blue which could be a Kingfisher anywhere near slow-moving water. Listen for the loud whistle that lets you know they are coming.

Feathered friends

At any time of year you will see birds around water. Some are migrants, so the species you see will vary with the season as well as with the habitat.

You may be lucky enough to see a Kingfisher, although these birds prefer fishing in slow-moving or still waters. Get to know their call, a strident whistle, which may give you a heads-up that they are coming. The bright electric-blue and orange streak that follows is unmistakable. You may spot nesting holes in an opposite bank, and if you wait near a branch hanging out over the water you may even be able to watch a Kingfisher fishing.

A Grey Heron lifting on creaky great pterodactyl wings is likely to be sighted. However, to watch one stalking in the water is to experience a masterclass in stillness and patience that you could only try and emulate.

Watch the rocks in fast upland streams and rivers and you may see Dippers, Grey Wagtails or Common Sandpipers. Dippers are dumpy brown and black birds with a big

Below: Be patient and stalk like a heron; you never know what wildlife encounters you may witness. This pterodactyl of the bird world has just caught a Grass Snake swimming across the lake.

white bib, and constantly bob or dip. They eat insect larvae and Freshwater Shrimps, and can dive and walk under water to get their prey. Grey Wagtails are actually a beautiful yellow and slate blue. Constantly wagging their long tails, they can be seen catching insects in flight over rocks and rivers. You may hear the three-note whistle of Common Sandpipers in the north and west of the UK.

Swallows, Swifts, House Martins and Sand Martins can be seen feeding over water during summer. You may even find a Sand Martin colony full of these lovely little brown and white-bottomed birds. They are one of the earliest migrants to return to our wetlands in spring. For a few short, wonderful weeks, you might hear Cuckoos anywhere in the country, or Nightingales in the south and east of England, particularly in reedbeds. Lapwings, Redshanks and Oystercatchers may number among the waders, and Sedge, Reed and Cetti's Warblers bring the Phragmites reeds to life with their songs. A Hobby or Marsh Harrier may pass overhead as it hunts.

In winter look for Snipe and Woodcock, and listen for the strange, pig-like squeals and grunts of the Water Rail. Wader and wildfowl levels increase with migratory birds, and over reedbeds you may see 'murmurations' of Starlings – the great displays that flocks of these birds make before roosting.

Above: Dumpy Dippers love fast-moving streams and can walk underwater.

Above: Spot Common Sandpipers 'teetering' up and down in the water.

Above: Marsh Harriers hunt low over reed beds.

Above: Water draws all manner of animals to it, including Muntjac and other deer; stay alert and look for other tracks and signs as you go.

Other mammals

If you are sitting or walking quietly while looking out for Otters, you are best placed to see other mammals that live in the water and its environs, as well as those that come down to the water to drink. Keep an eye out for deer, particularly Roe, Muntjac and Chinese Water deer, depending on where you are in the country. You may come across foraging Badgers or hunting Foxes on the banks. The chestnut streak of a Stoat with the black tip to its tail, a smaller Weasel, or even a Polecat or Mink may whizz past. These animals can be quite bold and may even investigate you!

Look for tiny Bank voles and even Harvest Mice in reedbeds. If you are really lucky you may spot the silvery shape of a swimming Water Shrew, or hear the distinctive 'plop' of a Water Vole entering the water. Though Water Voles are our fastest declining mammals, if present they can be relatively easy to watch, sitting on rafts of Water Crowfoot or weeds, munching stems, or ferrying across the river like buoyant scrubbing brushes.

On a summer's evening you can also look forward to some extremely close encounters with bats as they swoop over the surface of the water and around your head.

Above: The first sign of a Water Vole is often the 'plop' as it dives into water. If the habitat is good enough for Otters, they may be present.

Insects

In a good river habitat – and particularly in summer – you will be spoiled for insects to admire both above and below the water.

Butterflies and moths, damselflies and dragonflies patrolling on cellophane-crackly wings, and stoneflies and mayflies are just some of the insects you can see. Over Comfrey in summer you might see lots of brilliantly coloured Scarlet Tiger Moths, or even witness the magic of a mayfly hatching.

Beneath the water, or finely balanced on the meniscus of the water's surface, you might spot water boatmen, whirligig beetles, water scorpions or a Great Diving Beetle.

Above: Some spectacular insects such as this Southern Migrant Hawker abound by water as well as in it. Look for dragonflies and damselflies or the ephemeral glory of a mayfly hatch.

More wildlife

As you sit and wait get to know the plants, including trees, which make it all happen. Rushes, reeds and sedges add a unique atmosphere to a place, and a stunning array of flowering plants such as frothy Meadowsweet, tall Purple Loosestrife, Hemp-agrimony, Brooklime, Butterbur, Cottongrass and Yellow Flag Iris makes summer rivers idyllic places to spend some time. Learn to recognise poplar trees that constantly whisper of water, and damp-loving sallows, willows and Alders with their little purple cones that finches flock to in autumn.

Fish jumping for flies are mesmerising to watch, as are minnows in the shallows at your feet, or a great pike lurking in the shadows with its comically sinister, toothy grin. Polarising sunglasses can help you look into the water.

Look out, too, for frogs, newts and toads, or the thrill of a grass snake swimming across the water.

The truth is that if you sit still in a wonderful place for long enough, you will see things. Just do not get too distracted – you might miss your Otter.

Above: Meet the plants that make a habitat a home, whether it is summer colour in this Purple Loosestrife, or the sculptural beauty of *phragmites* reeds in winter.

Salty Dogs:
an Adaptation

Scottish Otters are often known by their Gaelic name, *dobhra*, or *beaste dubh* (black beast), and in Shetland they are known as the *dratsie*. The Otter's decline in Scotland mirrored what was happening in England and most of Wales, but (as in Ireland) it was not as acute, with a stronghold remaining in the cleanest waters in the north and west. Now, Otters have been recorded in every 10 sq km (4 sq miles) in Scotland, having returned to nine out of ten sites where they were present before the 1960s, making the most of the country's numerous burns, rivers and lochs.

However, Scotland is particularly known for its coastal-living, day-hunting Otters, the stars of wildlife films and documentaries, and something of a tourist attraction. It is thought that perhaps 50 per cent of Scottish Otters are coastal dwellers, feeding almost exclusively in the shallow, rocky seas around the wild and atmospheric coast and islands of western Scotland, the Outer Hebrides, Orkney and Shetland. In Shetland there is an Otter for every one of the 1,287km (800 miles) of coastline. It is the densest Otter population in Europe and nowhere else is quite like it.

The same but different

Living and feeding in the sea, these Otters are sometimes referred to as 'Sea Otters', which they are not – Sea Otters are a completely different species (see pages 18–29). These salty sea dogs are the very same species of Eurasian Otter that might frequent a Devon stream or a Suffolk marsh. Many of these riverine or freshwater Otters find their way, as rivers do, to the sea. They hunt on the seashore and in coastal waters as either part of their home range, or on

Above: A genuine Sea Otter *Enhydra lutris*, is an entirely different species to the Eurasian Otters living on the coast.

Opposite: River Otters have a taste for seafood. Half of Scotland's Otters live on the coast, hunting in daylight, making it the best place to see Otters in Britain.

exploratory hunting expeditions. Being a coastal-hunting Otter is not exclusive to these Scottish animals, yet they can be said to be ecologically distinct. They have a different way about them, and they live differently.

Seaweed and bathing pools

Scottish coastal Otters live on rocky, seaweedy shores that slope gradually into the sea. These shallow waters, coupled with large and small rock pools and forests of kelp, feed and harbour many types of fish and crustacean that are easily caught. Long, thin, sheltered lochs in the Western Isles reach like fingers into the sea, and around these coasts and islands there are also plenty of freshwater pools and lochs. These provide somewhere for the Otter to drink and fish when the sea is rough and stormy (although some individuals choose to hunt rabbits, voles and mice during bad weather). However, fresh water is also vital in keeping Otters' coats clean and functional: salt crystals from sea water build up in an Otter's fur, making it sticky and compromising its insulating properties. Otters need to visit 'bathing pools' approximately once or twice a day to wash themselves clean and reinsulate their pelts.

Below: These coastal Otters are not completely adapted to hunt at sea. They must regularly seek freshwater to drink and bathe in. Sticky salt crystal deposits from seawater can affect the efficiency of their pelts.

Time and tide (and seafood)

Coastal Otters still prefer bottom-feeding, eel-like fish similar to the prey of freshwater Otters, but the sea has an altogether different rhythm and timing that these fish react to. Here, Otter favourites such as eelpout, rockling, saithe and crabs are highly active at night, swimming out into deeper water beyond the diving reach of Otters. In the daytime, however, these fish mooch about in the shallows, under rocks and seaweed where they can be easily found. Other fish favourites such as sea scorpions, blennies and butterfish are active by day and night, but are much slower and can be found along the shallow sea bottom near where the other fish hide during the day. Almost all of these fish, however, become very active during the peak of high tide, so the Otters choose falling, rising or particularly low tides to feed in during the day, maximising their feeding success and expending the minimum amount of energy fighting the waves.

Above: Crab and eel-like fish are favourite prey, but instead of battling stormy seas in bad weather, Otters resort to land, hunting Rabbits, voles and mice.

Coastal ranges

Despite not being limited by two riverbanks, seashore territories are still linear. Otters here are simply foragers of the seabed rather than the riverbed, so as soon as the ocean gets too deep it is out of reach. Their seashore territory is rarely wider than a big river, and the extra space the ocean provides is an illusion – the animals here are still fiercely territorial. However, these shoreline waters are rich in feeding opportunities and Otters quite happily make lots of repeated, shallow and successful dives into the kelp forests within a small area, an activity known as patch fishing. Consequently, these areas can support many more Otters than inland ones do, and ranges can be as small as 4–5km (2½–3 miles) of coastline.

Above: Otters are foragers of the seabed, just as they are the riverbed, but cannot dive too deeply. Their territory is still linear, hugging the coastline.

Summer babies

Above: Coastal cubs are born and raised in holts with a sea view, often in old Rabbit burrows or small caves among rocks. With harsher winters to contend with, they are mostly born in summer.

Inland Otters do not have a set breeding period, as breeding does not need to coincide with the availability of any particular food source, and winters do not tend to be too severe. However, in Shetland and the north and west of Scotland, cubs are born mainly between May and August. From January to April food is far less abundant. Coastal Otters also tend to have smaller litters of cubs than those living on rivers, but the reason for this is not known.

Sea holts on the seashore

Most seaside Otter holts and dens are a little way in from the shore, where they can be excavated from sand dunes, peat or old Rabbit holes. The roots of the occasional windblown tree will be utilised, or a small gap or cave between rocks is sought out. Couches are made on beds of seaweed, or piles of sticks, driftwood, flotsam and jetsam. But Otters often sleep out in the open on the beach, curled up on rocks in the sun or on a patch of seaweed.

Seeing coastal Otters

Watching coastal Otters in Scotland is far easier than trying to spot one along the 40km (25 mile) length of a meandering river at dusk. There are more of them, they are out in the open during the day, their ranges are smaller and there are often people about 'in the know' who can help you. There is an Otter hide at Kylerhea, near the Glenelg Ferry, on Skye, for example, where wild Otters are almost reliably seen. Shetland, the Outer Hebrides, Skye and Mull are all good places to try – and wonderful to visit anyway. Even here, it still takes a lot of luck to spot an Otter.

Because of Otters' super-low profiles and seaweed-brown pelts, they are difficult to spot even in flat, calm

Below: The usual rules apply: dress warmly, keep a low profile, look twice and expect little. A horizontal, seaweed-coloured animal on a horizontal plane is difficult to spot, even in daylight.

Above: As riverine Otters go urban during the day, Scottish Otters are seen on the day shift at Shetland's busy oil terminal and from the car ferry.

waters, let alone in mildly choppy ones. Watch for the curve of an Otter's back as it dives, the tail breaching the surface and pointing wildly to the sky as it follows on with barely a splash. On land you may mistake several seabirds for an Otter, before glimpsing cubs following their mother like a brown Otter-skin cloak, rippling and waterfalling over the rocks behind her all the way down to the sea.

To watch Otters on such a horizontal plane, you will need to keep a low profile, too. It is best to sit down and wait on the wet, slippery, seaweedy rocks. If you do see an Otter and it dives, you have a chance to get closer, scotching forwards on your bottom as far as you dare, for the few seconds before it reappears and you must freeze again.

Otters' tolerance of people here is high – they are seen regularly in the waters around Shetland's Sullom Voe oil terminal, the largest in Europe. Some of the best sightings of Otters are from cars waiting to board the inter-island ferries. Excellent spaces for dens are nestled between the big boulders of the breakwaters.

Stars of the screen

Because of the relative ease with which Scottish coastal Otters can be watched, combined with their diurnal habits, they are probably the most studied and filmed Otters in Britain. They are valued, too, for their attraction to wildlife watchers and tourists, with many holidays, including walking, boat and Land Rover trips, on offer specifically to see them. Sandaig, a small archipelago of islands near Glenelg in the Scottish Highlands, is Gavin Maxwell's 'Camusfearna' of *Ring of Bright Water* fame and one of the most visited areas on the north-west coast.

Scotland's Big 5 is a recent partnership of organisations led by the Scottish government, Visit Scotland and Scottish Natural Heritage in a celebratory recognition of Scotland's wild heritage through its most iconic animals. Out of the Big 5 (the other four being the Golden Eagle, Red Squirrel, Harbour seal and Red deer), the Otter, 'the whiskered diver, the wild spirit of the waters' is considered the 'most-want-to-see'.

Above: Scotland's Otters are thriving and popular with hopeful visitors. Although tragedies still happen, signs like this Hebridean one slow the cars and keep visitor's eyes alert.

Below: Reclining on a couch of seaweed, Scotland's Otters are the most filmed and studied in Britain. They are celebrated as part of a rich wildlife heritage.

The Otter in Writing

Before the 1900s, Otter hunting was commonplace, because they were seen as a threat to human fish supplies. Otters were consequently not a popular choice for characters in literature at this time, although our Celtic heritage includes tales of Otter spirits and Otter kings. The Otter's ability to appear then disappear in the blink of an eye has lent itself to local myths and legends about black beasts and water hounds throughout the ages.

Accounts about Otters during the 20th century marked a shift in the way they were perceived and a rich seam of writing about Otters developed in the UK. Attitudes towards Otter hunting were changing, but still wild Otter numbers were decreasing. Several key authors wrote about Otters they had rescued or kept as pets, showing Otters in a more positive light. Through these stories and factual accounts about saving Otters or trying to live with them, novelists and naturalists recorded their observations of Otters and their relationships with individuals. At times writers 'became' the animals, imagining the world through their eyes. Literature played a significant role in raising the profile of Otters.

Today, Otters are popular and much loved. Having almost unwittingly wiped them out, we rediscovered them just before it was too late, returning them to their rightful place within our precious rivers and coastal waters. Otters have become a symbol of what the best kind of human effort and endeavour can achieve.

Above: Author of the book, known simply as *Tarka* (1927), Henry Williamson pricked consciences and challenged contemporary attitudes to Otters in Britain.

Opposite: A window in Overstrand Church, Norfolk depicts the 7th century legend of St Cuthbert and the Otters, one of the oldest pieces of writing about Otters viewed as anything other than 'pests'.

A children's tale

Kenneth Grahame's classic story *The Wind in the Willows* (1908) is the charming tale of riverside friends Mole, Rat, Badger and Toad and their adventures. Otter has a minor but important role. He is fierce and quick to defend, self-possessed, private, cheerful and independent. He has a particular sense of family, but also has the habit of appearing and disappearing at will, often in mid-sentence. Where Otters appear in children's stories, they often reveal recurrent 'ottery' traits.

Jill Tomlinson's *The Otter Who Wanted to Know* (1979) is about a Sea Otter with an insatiable curiosity that eventually leads her from peril to be rescued by humans. Mairi Hedderwick's beautiful picture book *The Utterly Otterlys* (2006) is about a family of Scottish Otters. Pa Otterley is restless and grumpy. His family have a strong sense of home life, but travel on Pa's whim to find a new abode.

In *Harry Potter and the Order of the Phoenix* (2003), Hermione Granger's Patronus, a magical guardian who takes the luminous, ghostly shape of the animal with whom they share the deepest affinity, is an Otter. Author J. K. Rowling has said that the character of Hermione was like an exaggerated form of herself when she was a child, and that the Otter was her favourite animal. In the film version of the book, the Otter Patronus is a Eurasian Otter.

A Cornish Otter

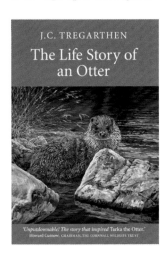

J.C. TREGARTHEN

The Life Story of
an Otter

'Unputdownable! The story that inspired Tarka the Otter.'
Howard Curnow, CHAIRMAN, THE CORNWALL WILDLIFE TRUST

Above: Tregarthen also wrote 'the lives' of Badger, Fox and Hare. His pioneering nature writing sought to describe their true lives, unsentimentally, from a wild animal's viewpoint.

In 1909, Cornish naturalist, fellow of the Zoological Society and schoolmaster John Coulson Tregarthen published his book *The Life Story of an Otter*. The book challenged prejudice against Otters and attempted to improve their public image.

The story is a close, intimate portrait of an animal he knew extremely well. Long before wildlife cameraman Charlie Hamilton James filmed Otters scenting prey underwater, Tregarthen suspected their ability to do so, writing that 'it may well be that the extraordinary powers of scent which enable the creature to detect the presence of fish in a stream or pond by sniffing the surface are called into play during immersion'. Through his writing he tried to get under the skin of the animal, behind its eyes and into the sense of it. He did not shy away from death, nor was he sentimental.

St Cuthbert and the Otters

St Cuthbert was a monk born in AD 635 who travelled to remote wild places to preach. He was well known for his remarkable affinity with animals, which extended to the introduction of what are considered to be the world's first laws protecting wildlife, preventing Farne Islanders from eating Eider ducks and other seabirds and their eggs. In a story recounted by the Venerable Bede (AD 673–735), St Cuthbert was observed in his habit of walking out to sea nightly to pray. When he left the water and knelt on the shore he was accompanied by a pair of Otters, who warmed his feet with their breath and dried him with their fur, rolling and playing around him. Statues and stained glass windows picture St Cuthbert with the Otters at his feet to this day.

Left St Cuthbert had a strong relationship with Otters and one not popular until we began to lose them in the 20th century. Now Otters flow easily into our imaginations and we dream of having an encounter with them.

Tarka the Otter

Henry Williamson was the author of many books, but perhaps his most famous book was about Otters, and he fully acknowledged Tregarthen's book as an inspiration. *Tarka the Otter: His Joyful Water-Life and Death in the Country of the Two Rivers* was first published in 1927. The two rivers are the Torridge and the Taw in Devon, and the book follows the two-year journey of Tarka from birth to death.

Williamson's writing style is vivid, with everything observed as if through the eyes of an Otter. The animals may have names, but they do not talk, and have only animal instincts and basic emotions such as fear, anger and pleasure.

Williamson was a restless man, troubled by his experiences in the Second World War and described by his daughter-in-law as difficult and artistic. He withdrew from people while researching *Tarka*, tracking animals in his wanderings and often sleeping rough.

Many acclaimed *Tarka*, including Thomas Hardy, John Galsworthy, T. E. Lawrence and Edward Thomas, for its intimacy and sense of place. Rachel Carson, author of *Silent Spring* (1962), cited the book as one of her biggest influences. It changed the way people thought about Otters and highlighted the brutality of Otter hunting.

Decades later you can cycle the Tarka Trail and follow the real-life route of Tarka's journey, where several of the holts Williamson wrote about in 1927 are still being used by Otters.

Above: A classic: the First edition cover of *Tarka*, by artist Hester Sainsbury. With an estimated 4 million copies sold, it has never been out-of-print.

Right: The long distance, off-road Tarka Trail is one of North Devon's famous visitor attractions, passing through countryside described in the book. It takes in such real-life literary landmarks as Beam Weir (pictured) where Tarka played.

Female influences

In the same year that *Tarka* was published Frances Pitt, a young foxhunting woman and animal psychology student, kept three Otters, Moses, Aaron and Tom, to study as pets. At the time river bailiffs and gamekeepers still shot Otters on sight. She attempted to challenge misconceptions and concluded that Otters were second only to dogs in their intelligence. The subtitle of her book *Moses, My Otter* (1927) names the Otters and their friend 'Tiny the Terrier', highlighting an understanding between the animals.

In the 1940s, Phyllis Kelway was out Otter hunting, which she described in *The Otter Book* (1944). She saw an Otter cub frantically trying to swim upstream against a flow that was pulling the cub towards a deadly weir. Kelway leapt in, risking her own life, and grabbed the Otter by its tail. She never hunted again. Naming the cub Juggles, she set about changing the reputation of wild Otters with the intention that, 'if [Juggles] lived, she should be the means of saving the lives of many other otters'. By getting to know Juggles as she did through her story, Kelway hoped people would be moved to stop hunting Otters.

Below: Daphne Neville's Asian Small-clawed Otters (like this pair) wriggled out of her outsized jumper, to entertain, educate, and inspire generations, becoming important ambassadors for British Otters.

Over the following decades, several authors followed these women's leads, writing about Otters they had rescued, studied or obtained as pets, and by so doing achieving much to set Otters firmly in a new light. Notably, in the 1980s and 1990s former actress Daphne Neville did much to raise awareness of the plight of otters worldwide with her pet Asian Small-clawed Otters. She began to travel the country, speaking at shows and events and to countless schoolchildren. Her children's book *Bee, a Particular Otter* (1982) recounts her Otter Bee's real-life adventures with royalty as well as her brushes with the law.

Otter protectors

Of course, along with *Tarka*, the most significant of the books to turn the British people into emphatic Otter lovers and protectors was Gavin Maxwell's *Ring of Bright Water*, published in 1960 and followed by *The Rocks Remain* (1963) and *Raven Seek Thy Brother* (1968).

Ring of Bright Water tells the autobiographical story of how Maxwell brought a Smooth-coated Otter back from Iraq, named it Mijbil and raised it as a pet and a replacement for his much-loved spaniel. Maxwell realised just how much an Otter misses water if kept from it when, in Basra, he put the pining Otter in the bath. '…for an hour he went wild with joy in the water, plunging and rolling in it, shooting up and down the length of the bath underwater'.

The story follows Maxwell and Mij from a flat in London, where Mij caught eels in the bath and slept with Maxwell, to the cottage ringed by water on the wild Scottish shores of Sandaig (Camusfearna as it was known in the book). Mij walked on a lead, learned to turn on taps, played with balls and came when called. But, he also caused havoc when bored; Maxwell quickly acknowledged that Otters were not dogs.

His story recounts many touching incidents with Otters. When travelling with Mij in the first-class sleeper train from London to Scotland, Mij had to be 'tickled' away from pulling the train's emergency cord and later slept with his head on the pillow with Maxwell, his 'arms' over the sheets in an almost human way.

Other pet Otters followed and in *Raven Seek Thy Brother* (1968) the Otters Edal and Teko had to be kept in near zoo-like conditions to prevent them making seemingly unprovoked attacks on people. Keeping Otters like this deeply saddened Maxwell. When Edal died tragically in a fire that also burnt down the house at Camusfearna, he left and did not return.

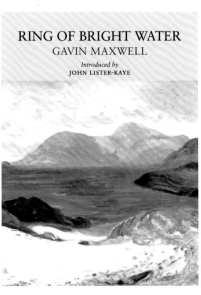

RING OF BRIGHT WATER
GAVIN MAXWELL
Introduced by
JOHN LISTER-KAYE

Above: Maxwell's autobiography of living with his pet Otters entertained and moved many readers and still does today; it is responsible for inspiring a generation of conservationists.

THE WHITE ISLAND

John Lister-Kaye

The ripples that spread after the publication of *Ring of Bright Water* travelled far. Maxwell employed a succession of Otter-keepers – one of whom, Terry Nutkins, went on to inspire a generation of nature lovers as a TV presenter of wildlife shows. Edal famously bit off the tips of two of Nutkins' fingers. Another close friend with whom Maxwell shared Camusfearna and had a difficult relationship with was the poet Kathleen Raine. Maxwell was the unrequited love of her life, yet he ultimately held her responsible for much of the 'bad luck' that

Left: The renowned Scottish naturalist and nature writer, John Lister-Kaye, finished *The Ring* story after Maxwell's sudden death in 1969 ended their joint conservation project.

Below: Edal's memorial at Sandaig (Camusfearna). Stories about Otters rarely ended well, but ultimately, they fired our imaginations and no doubt saved the wild Otter from extinction.

EDAL, THE OTTER OF RING OF BRIGHT WATER 1958 – 1968 Whatever joy she gave to you give back to Nature. GAVIN MAXWELL

dogged his and the Otters' life. Another employee, John Lister-Kaye (later Sir John), went on to finish Maxwell's story in *The White Island* (1972). He is one of Scotland's best-known naturalists and nature writers, and set up the Aigas Field Centre in the Highlands in Maxwell's memory in 1976.

A film loosely based on *Ring of Bright Water*, released in 1969, did much to popularise Otters and bring the idea of them closer to people's hearts (the shocking scene of Mij's senseless death was particularly influential). It also brought *Tarka* to the attention of a new generation, ensuring the book's place as a classic. It was brilliantly made into a film in 1979. Henry Williamson died on the day the last scene – of Tarka's death at the jaws of the pied hound Deadlock – was finished.

Williamson and Maxwell were both complicated, creative and flawed individuals; they were restless, Otter-like loners who nevertheless changed our attitude to the Otter in Britain.

Below: The 1969 film *Ring of Bright Water*, starring Bill Travers and Virginia McKenna, was sentimental, but did not shy away from its brutal, tragic end.

Facts, film and family

Many reference books written about Otters convey a tone of wistfulness as well as a sense of humour from their authors – particularly Paul Chanin and James Williams. Paul Chanin's tips on watching Otters involve waiting by any Devon bridge for a fortnight, whereby an Otter will pass by at some point. James Williams admits that the animals 'cast a powerful and compelling spell: those that try to enter their mysterious realm invariably become "besottered".'

In the 21st century, writing about the trials of filming wild Otters has replaced writing about the trials of keeping pet Otters. Philippa Forrester's book *The River: A Love Story* (2004) beautifully charts how Otters, as well as Kingfishers, became part of her family's life during filming for *My Halcyon River* (2004) and *Halcyon River Diaries* (2010), filmed by her wildlife-cameraman husband Charlie Hamilton James on the river outside their home. Halcyon was the name of the Kingfisher in *Tarka*. Another wildlife cameraman, Simon King, filmed and wrote about Scottish coastal Otters in his *Shetland Diaries* (2010).

Perhaps the two best-known poems written about Otters are by two of the most eminent poets of the 20th and 21st centuries, Ted Hughes and Seamus Heaney. Nature is an important influence on them both. Heaney's *The Otter* was a love letter to his wife. While Ted Hughes (1930–1998) admitted to being 'profoundly affected' by reading *Tarka* and this is clearly reflected in his poem *An Otter* (1960).

Other modern poets have taken on the mantle and, more recently in 2012, poet Miriam Darlington's prose-poem *Otter Country* carries on the tradition of beautiful writing about Otters in Britain, while also breaking new ground. In it the 21st-century Otter is still elusive, but recovering. Darlington's book looks anew at a country where a wild animal we thought lost has returned to us; an animal that remains mysterious and elusive, yet surprises us with its adaptability, its appearances and non-appearances.

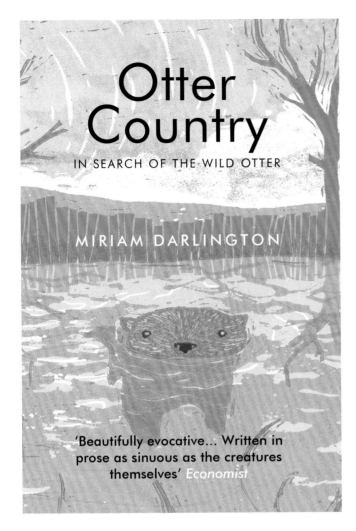

Otter Country

IN SEARCH OF THE WILD OTTER

MIRIAM DARLINGTON

'Beautifully evocative... Written in prose as sinuous as the creatures themselves' *Economist*

Left: Miriam Darlington's prose-poem *Otter Country* covers the loss and return of the Otter.

Otters are physically more real and present today than they have been for half a century, but we are also more aware of them and excited by the thought of their presence. Otters today have a place in the public imagination and consciousness like never before. The story of the Otter is far from over and, in many ways, is beginning afresh. As Paul Hyland writes in his poem *Otter Returns* (2001) Britain has once again become 'home to a species mislaid', the Otter re-entering our literature and imagination from the tributary of its recovery.

Glossary

Carnassial, the sheering cheek teeth of a true carnivore, e.g. a Weasel

Couch, a temporary 'campsite', a sleeping or resting place for an otter

Fieldcraft, the art of watching and tracking wildlife

Herbicide, a chemical used to kill weed 'pests'

Holt, an otter's den, often below ground or under the roots of trees

Hover, a less permanent resting place for an otter, above ground

Lutrine, otter-like, from the animal's scientific name, *Lutra lutra*

Metabolism, the speed at which an animal digests and converts food to energy

Mustelid, belonging to the weasel family

Pelt, an otter's skin and fur

Pesticide, a chemical used to kill insect 'pests'

Piscivore, a fish-eater

Raft, a group of otters swimming closely together or a family of otters

Reintroduction, where captive bred animals are put back into the wild

River Bailiff, a person employed to manage a river for commercial fishing

Riverine/ Riparian, relating to a river, its banks and habitat

Romp, a group of otters playing

Rudder, an otter's thick muscular tail

Seal, an otter's pawprint

Spate, a river running high and fast with rain or flood water

Spraint, otter poo and a scented messaging system combined

Tributary, a stream that flows into a larger stream or river

Vermin, an animal considered a pest and therefore often allowed to be killed

Vibrissae, the highly sensitised whiskers on an otter's muzzle

Further Reading

Fiction

Chaffe, David, *Stormforce; an Otter's Tale* (Stormforce Publications, 1999), part fiction.

Grahame, Kenneth, *The Wind in the Willows* (Methuen, 1908).

MacCaskill, Bridget, *A Private Sort of Life* (Whittles Publishing, 2002), part fiction.

Maxwell, Gavin, *Ring of Bright Water* (Longmans, 1960).

The Rocks Remain (Longmans, 1963).

Raven Seek Thy Brother (Longmans, 1969).

Rowling, J. K., *Harry Potter and the Order of the Phoenix* (Bloomsbury, 2003).

Tregarthen, J. C., *The Life Story of an Otter* (Read Books Design 1909, reprinted 2011).

Williamson, Henry, *Tarka the Otter: His Joyful Water-Life and Death in the Country of the Two Rivers* (G. P. Putnam's Sons, 1927).

Non-fiction

Allen, Daniel, *Otter* (Reaktion, 2010).

Chanin, Paul, *Otters* (Whittet, 1993, revised 2013).

Darlington, Miriam, *Otter Country. In Search of the Wild Otter* (Granta, 2012).

Forrester, Philippa, *The River: A Love Story* (Orion, 2004).

Forrester, Philippa & Charlie Hamilton James, *Halcyon River Diaries* (Preface, 2010).

Kelway, Phyllis, *The Otter Book* (Collins & Co., 1944).

King, Simon, *Shetland Diaries* (Hodder & Stoughton, 2010).

Kruuk, Hans, *Otters: Ecology, Behaviour and Conservation* (Oxford, 2006).

Lister-Kaye, John, *The White Island* (Longman, 1972).

MacCaskill, Bridget, *On The Swirl of the Tide* (Luath Press, 1992).

Neal, Ernest, *Topsy & Turvy, My Two Otters* (Heinemann, 1961).

Pitt, Frances, *Moses, My Otter* (Arrowsmith, 1927).

Talbot, Peter, *Tarka and Me* (Kindle edition, Pambazuku, 2011).

Tulloch, Bobby, *Otters*

(Colin Baxter Photography, 1994).
Walton, Izaak, *The Compleat Angler* (1653; Oxford University Press, 1982).
Wayre, Philip, *The River People* (Collins & Harvill Press, 1976).
The Private Life of the Otter (Book Club Associates, 1979).
Williams, James, *The Otter Among Us* (Tiercel, 2000).
Williams, James, *The Otter* (Merlin Unwin, 2010).
Yalden, Derek & Stephen Harris, ed., *Mammals of the British Isles: Handbook*, 4th ed. (The Mammal Society, 2008).
Yoxon, Paul & Grace, *Echoes of Camusfearna* (Findhorn, 1997).

Poetry

Heaney, Seamus, *The Otter from Field Work* (Faber & Faber, 1979).
Hughes, Ted *An Otter from Lupercal* (Faber & Faber, London, 1960).
Oswald, Alice, *Dart* (Faber & Faber, 2002).

For young children

Ezra, Mark, illustrated by Gavin Rowe, *The Hungry Otter* (Litter Tiger Press, 1997).
Goldsmith, John, *Tarkina the Otter* (Pelham, 1981).
Hedderwick, Mairi, *The Utterly Otterleys* (Hodder, 2006).
Neville, Daphne & Ken Jackson, *Bee, a Particular Otter* (Windmill, 1982, reprinted 2012).
Tomlinson, Jill, *The Otter Who Wanted to Know* (Methuen, 1979).

Resources

RSPB www.rspb.org.uk
The Wildlife Trusts www.wildlifetrusts.org
The Vincent Wildlife Trust www.vwt.org.uk
The Environment Agency www.environment-agency.gov.uk
Natural England www.naturalengland.org.uk/ourwork/regulation/wildlife/species/otters.aspx
The Mammal Society www.mammal.org.uk
International Union for Conservation of Nature (IUCN) www.otterspecialistgroup.org
International Otter Survival Fund www.otter.org
Cardiff University Otter Project www.otterproject.cf.ac.uk
Somerset Otter Group www.somersetottergroup.org.uk
Sustainable Eels Project www.sustainableeelgroup.com
Otter Joy www.otterjoy.com
Amblonyx Otter www.amblonyx.com

Otter Sanctuaries

British Wildlife Centre, Surrey, RH7 6LF www.britishwildlifecentre.co.uk
New Forest Otter, Owl and Wildlife Park, Hampshire, SO40 4UH www.ottersandowls.co.uk
The Chestnut Centre, Derbyshire, SK23 0PE www.chestnutcentre.co.uk
Buckfast Butterfly Farm and Dartmoor Otter Sanctuary, Devon, TQ11 0DZ www.ottersandbutterflies.co.uk
Tamar Otter and Wildlife Centre, Cornwall, PL15 8GW www.tamarotters.co.uk
The Scottish Sealife Sanctuary, Argyll PA37 1SE www.sealsanctuary.co.uk

Image credits

Bloomsbury Publishers would like to thank the following for providing photographs and for permission to produce copyright material. While every effort has been made to trace and acknowledge all copyright holders, we would like to apologise for any errors or omissions and invite readers to inform us so that corrections can be made in any future editions of the book.

Photographs
Key t=top; l=left; r=right; tl=top left; tcl=top centre left; tc=top centre; tcr=top centre right; tr=top right; cl=centre left; c=centre; cr=centre right; b=bottom; bl=bottom left; bcl=bottom centre left; bc=bottom centre; bcr=bottom centre right; br=bottom right

AL = Alamy; FL=FLPA; G = Getty Images; NP = Nature Picture Library; RS = RSPB Images; SS = Shutterstock

Front Cover t Paul Hobson/FL; b Mark Bridger/SS; **1** Stephan Morris/SS; **3** Flip De Nooyer/FL; **Back Cover** t David Tipling/RS; b David Kjaer/RS; **4** Paul Hobson/FL; **5** Foto Natura Stock/FL; **6** Robin Chittenden/FL; **7** Ingo Arndt/FL; **8** David Tipling/RS; **9** David Tipling/RS; **10** /Imagebroker/FLPA; **11** Hans Dieter Brandl/FL; **12** Rod Teasdale; **13** Imagebroker, Marcus Siebert/FL; **14** Malcom Schuyl/FL **15**tr Roel Hoeve/FL; cr Eduard Kyslynskyy/SS; bcr David Kjaer/RS; br Mark Hamblin/RS; **16**tl David Kjaer/RS; cl Laurie Campbell/RS; bcl Jagodka/SS; bl Mike Lane/RS; **17** Robin

Chittenden/FL; **18** poeticpenguin/SS; **19** Jeroen Stel/RS; **20** bluehand/SS; **21** Pierre Torset/FL; **22** Suzi Eszterhas/FL; **23** Joerg Reuther/FL; **24**t Pete Oxford/FL; b Danny Ellinger/FL; **25** Pete Oxford/FL; **26** Edward Myles/FL; **27**t Kevin Schafer/FL; b Frans Lanting/FL; **28** Ingo Arndt/FL; **29** Julie Dando/Fluke Art; **30** David Kjaer/RS; **31** David Hughes/SS; **32**tc Ivonne Wierink/SS; tl Matthijs Wetterauw/SS; b Martin Fowler/SS; **33**t /Imagebroker/FLPA; b Michael Durnham/FL; **34** S Charlie Brown/FL; **35**t Emi/SS; b David Tipling/FL; **36** Ingo Schulz/FL; **37**t Gary K Smith/FL; b Mike Lane/RS; **38**t Paul Hobson/FL; b Neil Hardwick; **39** Michael Durham/FL; **40** Jack Chapman/FL; **41** David Tipling/RS; **42** Phil McLean/FL; **43** Derek Middleton/FL; **44**t Jack Chapman/FL; b Edwin Geisbers/NP; **45**t Dickie Duckett/FL; b Ernie Janes/RS; **46** Peter Weimann/G; **47** Ingo Arndt/FL; **48**t Michael Durham/FL; b Robin Chittenden/FL; **49** Frederic Feve/FL; **50** Richard Revels/RS; **51** David Tipling/FL; **52** David Tipling/FL; **53** Nicole Duplaix/G; **54** FL Jack Chapman; **55** David Tipling/G; **56** Mark Hamblin/RS; **57**c Paul Hobson/FL; b Yuriy Kulik/SS; **58** Emanuele Biggi/FL; **59**t Solvin Zankl/NP; b Michael Durham/FL; **60** Niall Benvie/FL; **61**t Jules Cox/FL; b David Tipling/RS; **62** David Burton/FL; **63** Marcus Sierbert/FL; **64**t Ramona Ritcher/FL; b Antonio Abrignani/SS; **65** David Fowler/SS; **66** Mike Lane/FL; **67** David Tipling/RS; **68** Nigel Cattlin/FL; **69** Paul Sawer/FL; **70** Viktoriya Field/SS; **71** Sue Kennedy/RS; **72** Paul Hobson/FL; **73** David Tipling/RS; **74** © Crown Copyright Forestry Commission; **75** Simon Litten/FL; **76** Roger Tidman/FL; **77** David Tipling/RS; **78** Imagebroker/FL; **79** Maik Blume/FL; **80**

Paul Sawer/FL; **81** Andy Gehrig/G; **82** David Woolfenden/SS; **83t** Frederic Desmette/FL; b Picavet/G; **84** Martin Fowler/SS; **86** Elliott Neep/FL; **87** Imagebroker/FL; **88** Michael Krabs/FL; **89** Jack Chapman/FL; **90** Michael Durham/FL; **91** Flip Nicklin/FL; **92** Michael Durham/FL; **93tl** Stephen Reese/SS; tr Dolnikov/SS; cr Michael Callan/FL; cl Keith Tarrier/SS; br Michael Durham/FL; **94t** Fabio Pupin/FL; b Phil McLean/FL; **95** Roger Tidman/FL; **96** Michael Durham/FL; **97** David Tipling/RS; **98** John Eveson/FL; **99** Malcolm Schuyl/FL; **100t** Maria Breuer/FL; b Paul Sawer/FL; **101t** Mike Lane/FL; c Jonathan Lhoir/FL; b Michael Durham/FL; **102t** Mike Lane/FL; b Jules Cox/ FL; **103t** James Lowen/FL; b Imagebroker/FL; **104** Jack Chapman/FL; **105** Donald M Jones/FL; **106** Michael Durham/FL; **107t** Michael Durham/FL; b Konrad Borkowski/FL; **108** Sylvain Cordier/FL; **109** Elliott Neep/FL; **110** Chris Gormersall/AL; **111t** Derek Middleton/FL; b Michael Durham/FL; **112** Rex Harris; **113** Getty Images; **114** J C Tregarthen/Cornwall Editions Ltd; **115** Dave Webster; **116** Culture Club/G; **117** Rod Teasdale; **118** SH Grant Glendinning; **119** Winifred Nicholson/Little Toller Books; **120t** Sir John Lister-Kaye/Aigas Books; b Duncan Astbury/AL; **121** G Michael Ochs; **123** Kelly Dyson/Granta Books; **126** FL Roger Tidman.

Tailpiece

Otters have evolved to diversify, adapt and deal with flooding, harsh winters and drought, but with an ever-increasing human population and the myriad pressures that brings to bear on our environment, the Otters that have returned to our waterways, seashores and wetlands will have to be more resourceful than ever.

Localised pollution and accidents still happen and insidious pollution from new drugs and chemicals that we barely understand still enters our waters. Habitat is still under threat. Valuable research on Otters, that also shines a light on our own health, is threatened by funding cuts. To the modern Otter, our rivers may be cleaner, but our roads are more deadly. And anglers, fish farmers and Otters will have to learn how to live together.

If an Otter were a more static creature, these pressures might prove catastrophic. But Otters are intelligent, curious, highly developed animals with individual, inquisitive and independent traits. All this can only help them negotiate the challenges of the future. This most ancient of beasts is still evolving, and they are more resilient, more tolerant, more adaptable and less shy than we ever knew. And we are on their side.

But our wildlife needs us to protect and conserve it like never before. The return of the Otter proves we can, and it demonstrates just how much we want and need our wild creatures in our lives. We have learnt much from the Otter and in return have been given a rare second chance to get it right.

Acknowledgements

I would like to thank Mum and Dad, and Roger and Sue Chester, the community of Inkpen village and school, and the lovely staff at Newbury Library (in particular Diane and Helen) for their support. Heartfelt thanks to Sarah Evans for her wonderful listening ear, sound advice and joie de vivre, and to Derek Niemann for his encouragement, general mentorship and the wise quote at the beginning of this book. Thank you to Krystyna Mayer, and especially to Alice Ward, Jasmine Parker and Julie Bailey at Bloomsbury. Special thanks to James Sadler – with whom I saw my first wild Otter – for lending and imparting his fieldcraft magic on all things wild. And of course, huge thanks to the romp and the raft that is my family: my children, Billy, Evie and Rosie, whose excitement in this project and faith in me never waned and who are still as enthusiastic about mud and wildlife as ever, and to Martin, for his unconditional love, understanding and enduring belief in me.

Index

INDEX

agro-chemicals 68–9,
70–1
anal scent glands 13,
22, 59
ancestry 12
appearance 6, 8–11

Badger 14, 15, 25, 46, 53,
93, 97, 102
bedding 39, 52–3
Blackbird 90
blennies 107
body shape 7, 13, 41–2
bream 34
breeding 51–3
bubble-blowing 48
bullhead 34
butterfish 107

calls 11, 25, 56, 90
carp 34, 36
Carson, Rachel 69, 116
cats 13, 93
chub 33
chutes 96
claws 19, 26
coastal Otters 22–3, 34,
105–7
crabs 11, 34, 48, 107
crayfish 34, 36, 97
crayfish, signal 35
cubs 52, 53, 94, 97, 108
independence 57
role of play 55–6
Cuckoo 101
curiosity 55

DDT 68, 70
deer 102
diet 11, 33–5, 97
Dipper 56, 100–1
diving Otters 43
dogs 13, 43, 49, 93

ears 6, 41, 42
eel 33, 34, 35, 48, 69,
71, 94
eelpout 34, 107
endangered species 19,
27, 28, 65, 66
eyes 6, 8, 41, 42, 46

Falcon, Peregrine 67,
68, 69
feet 7, 13, 42, 43, 49
females 7, 11, 51, 52–3
Ferrets 13, 14, 16
fieldcraft 92–3
fighting 57, 59
fish 11, 32, 33–4, 48, 94,
97, 103, 106

fish farms 37
fishing with Otters 21
food chains 71
Fox 39, 93, 97, 102
frogs 11, 34, 94, 97, 103
fur 13, 22–3, 25, 28,
44–5, 63

gait 49
game birds 67, 68
garden ponds 37
gestation 52
grass snake 35, 103
grooming 45
gudgeon 34

habitat 31–2
habitat loss 19, 20, 28
Heron, Grey 97, 100
holts 39, 52–3, 108
Honey Bee 71
hovers 39
hunting habits 36–7
hunting Otters 19, 20, 28,
63, 64–5, 116

insect life 103

Kingfisher 56, 100
Kite, Red 64

Lapwing 101
latrines 25
legs 7, 13
lifespan 11
loach, stone 34

males 7, 11, 51, 52
Marten, Pine 13, 14, 16
Martin, Sand 101
Maxwell, Gavin 111,
119–21
Meerkat 11
metabolism 13, 36
mice 12, 35, 102, 106
Mink 8, 16, 37, 65, 69,
90, 93, 95, 102
minnows 33
Mionictis 12
Mole, Star-Nosed 48
Moorhen 90
moose 14
mussels 34
Mustelidae 7, 11, 12, 13,
14, 22
British mustelids 15–16

names 7–8, 13, 105
newts 35, 103
Nightingale 101
nose 41, 42

nostrils 8, 42

Otter, African Clawless 26
Asian Small-clawed
19–20, 25
Congo Clawless 26
Otter, Eurasian 6, 15, 22,
26, 51, 56
range 28–9
Otter, Giant 13, 24–5
Hairy-nosed 27
Marine 27
Neotropical 27
North American 26
Otter, Sea 14, 22–3, 105
use of rocks as tools 23
Otter, Smooth-coated 21
Southern River 27
Spotted-necked 26
Oystercatcher 101

paths 95, 96
pawprints 92, 93
paws 19–20
pelt 44–5
perch 33
pike 34, 103
plant life 103
playing 51, 52, 55
Polecat 13, 14, 16, 102
Polecat-ferrets 16
pollack 34
pollution 19, 20, 23, 28,
68–9, 70–1
population decline 19, 27,
68–9, 70–1
'porpoising' 43

Rabbit 11, 14, 34, 39, 52,
106, 108
Rail, Water 101
Redshank 101
remains of meals 97
road deaths 19, 61
rockling 34, 107

saithe 34, 107
salmon 34
Sandpiper, Common 100,
101
scent marking 13, 58–60
sea scorpions 107
shrew, water 48, 102
sight 46
size 7, 11, 19
sleeping habits 38–9, 108
slides 56, 95, 96
smell 46, 48
Snipe 101
social habits 26, 51–2
spraints 13, 34, 35,

58–60, 94–5
'Otter seats' 95
St Cuthbert and the Otters
115
Starling 101
sticklebacks 33, 36
Stoat 14, 15, 102
Swallow 101
Swift 101
swimming Otters 8,
41–2, 54

tails 7, 8, 11, 41, 90, 110
teeth 48
tench 34
territories 13, 26, 38, 51,
57, 107
toads 34, 97, 103
toes 7, 13, 19, 26
tracking Otters 92–3
trail cameras 96
trapping 63
trout 34, 36, 37
tunnels 95, 96

vermin 63, 64
Vole, Water 35, 37, 102
voles 12, 35, 102, 106

Wagtail, Grey 56, 100,
101
Walton, Izaak 65
Warbler, Cetti's 101
Reed 101
Sedge 101
watching Otters 17,
87–88, 98–99
capturing the evidence
96
coastal habitats 109–11
fieldcraft and tracking
92–3
hints and tips 89
identification skills
90–1
where to find Otters 91

waterbirds 11, 34, 56, 95,
100–1
weaning 53
Weasel 7, 13, 15, 102
weight 11
whiskers 6, 47–8
Williamson, Henry 116,
121
wolverine 14
Woodcock 101
Woodpigeon 67, 69
writing about Otters
113–23